Environmental Aspects of Textile Dyeing

Environmental Aspects of Textile Dyeing

Editor

Jamshed A Khan

scitus academics

Environmental Aspects of Textile Dyeing
Edited by **Jamshed A Khan**

ISBN: 978-1-68117-459-4
Library of Congress Control Number: 2016937635

SCITUS Academics LLC,
www.scitusacademics.com
Box No. 4766, 616 Corporate Way,
Suite 2, Valley Cottage,
NY 10989

Printed in United States of America on Acid Free Paper

Preface

Textile colors enhance our environment and bring color to life. The current color range was developed to withstand environmental effects such as exposure to light and degradation of water. However, it is the industry's responsibility to apply coloring to textiles to ensure that the potential for harm to the environment is minimized, for example through residues in waste streams and to the consumer. In the textile industry, up to 200,000 tons of these colors are lost to effluents each year during coloring and finishing operations due to the inefficiency of the coloring process. Unfortunately, most of these colors, due to their high light stability, temperature, water, detergents, chemicals, soap and other parameters such as bleach and suddenness, escape conventional treatment processes for waste water and persist in environment. In addition, anti-microbial agents that resist biological degradation are often used in textile production, particularly in natural fibers such as cotton. The synthetic origin and complex aromatic structure of these agents makes them more biodegradable. The textile industry consumes significant amounts of water mainly used in the coloring and finishing operations of the plants in its manufacturing processes. Textile waste water is classified as the most polluting of all industrial sectors, taking into account both the volume produced and the composition of the effluent. This book concludes with chapters on the reduction of pollution and waste, reuse of spent dyebath, chemical treatment of dye effluent and biotechnological treatment of dye effluents. This book is a valuable tool for manufacturer's development of a sustainable industry.

Table of Contents

Chapter 1

Textile Dyes: Dyeing Process and Environmental Impact

Farah Maria Drumond Chequer[1], Gisele Augusto Rodrigues de Oliveira[1], Elisa Raquel Anastácio Ferraz[1], Juliano Carvalho Cardoso[2], Maria Valnice Boldrin Zanoni[2] and Danielle Palma de Oliveira[1]

[1] USP, Department of Clinical, Toxicological and Bromatological Analyses, Faculty of Pharmaceutical Sciences at Ribeirão Preto, University of São Paulo, Ribeirão Preto, SP, Brazil

[2] UNESP, Department of Analytical Chemistry, Institute of Chemistry at Araraquara, State of São Paulo University Júlio de Mesquita Filho, Araraquara, SP, Brazil

1. INTRODUCTION

Dyes may be defined as substances that, when applied to a substrate provide color by a process that alters, at least temporarily, any crystal structure of the colored substances [1,2]. Such substances with considerable coloring capacity are widely employed in the textile, pharmaceutical, food, cosmetics, plastics, photographic and paper industries [3,4]. The dyes can adhere to compatible surfaces by solution, by forming covalent bond or complexes with salts or metals, by physical adsorption or by mechanical retention [1,2]. Dyes are classified according to their application and chemical structure, and are composed of a group of atoms known as chromophores, responsible for the dye color. These chromophore-containing centers are based on diverse functional groups, such as azo, anthraquinone, methine, nitro, arilmethane, carbonyl and others. In addition, electrons withdrawing or donating substituents so as to generate or intensify the color of the chromophores are denominated as auxochromes. The most common auxochromes are amine, carboxyl, sulfonate and hydroxyl [5-7].

It is estimated that over 10,000 different dyes and pigments are used industrially and over 7 x 10^5 tons of synthetic dyes are annually produced worldwide [3,8,9]. Textile materials can be dyed using batch, continuous or semi-continuous processes. The kind of process used depends on many characteristics including type of material as such fiber, yarn, fabric, fabric construction and garment, as also the generic type of fiber, size of dye lots and quality requirements in the dyed fabric. Among these processes, the batch process is the most common method used to dye textile materials [10].

In the textile industry, up to 200,000 tons of these dyes are lost to effluents every year during the dyeing and finishing operations, due to the inefficiency of the dyeing process [9]. Unfortunately, most of these dyes escape conventional wastewater treatment processes and persist in the environment as a result of their high stability to light, temperature, water, detergents, chemicals, soap and other parameters such as bleach and perspiration [11]. In addition, anti-microbial agents resistant to biological degradation are frequently used in the manufacture of textiles, particularly for natural fibers such as cotton [11,12]. The synthetic origin and complex aromatic structure of these agents make them more recalcitrant to biodegradation [13,14]. However, environmental legislation obliges industries to eliminate color from their dye-containing effluents, before disposal into water bodies [9,12].

The textile industry consumes a substantial amount of water in its manufacturing processes used mainly in the dyeing and finishing operations of the plants. The wastewater from textile plants is classified as the most polluting of all the industrial sectors, considering the volume generated as well as the effluent composition [15-17]. In addition, the increased demand for textile products and the proportional increase in their production, and the use of synthetic dyes have together contributed to dye wastewater becoming one of the substantial sources of severe pollution problems in current times [6,9].

Textile wastewaters are characterized by extreme fluctuations in many parameters such as chemical oxygen demand (COD), biochemical oxygen demand (BOD), pH, color and salinity. The composition of the wastewater will depend on the different organic-based compounds, chemicals and dyes used in the dry and wet-processing steps [6,18]. Recalcitrant organic, colored, toxicant, surfactant and chlorinated compounds and salts are the main pollutants in textile effluents [17].

In addition, the effects caused by other pollutants in textile wastewater, and the presence of very small amounts of dyes (<1 mg/L for some dyes) in the water, which are nevertheless highly visible, seriously affects the aesthetic quality and transparency of water bodies such as lakes, rivers and others, leading to damage to the aquatic environment [19,20].

During the dyeing process it has been estimated that the losses of colorants to the environment can reach 10–50% [13,14,17,21,22]. It is noteworthy that some dyes are highly toxic and mutagenic, and also decrease light penetration and photosynthetic activity, causing oxygen deficiency and limiting downstream beneficial uses such as recreation, drinking water and irrigation [13,14,23]

With respect to the number and production volumes, azo dyes are the largest group of colorants, constituting 60-70% of all organic dyes produced in the world [2,24]. The success of azo dyes is due to the their ease and cost effectiveness for synthesis as compared to natural dyes, and also their great structural diversity, high molar extinction coefficient, and medium-to-high fastness properties in relation to light as well as to wetness [2,25]. They have a wide range of applications in the textile, pharmaceutical and cosmetic industries, and are also used in food, paper, leather and paints [26,27]. However, some azo dyes can show toxic effects, especially carcinogenic and mutagenic events [27,28].

The toxic effects of the azo dyes may result from the direct action of the agent itself or of the aryl amine derivatives generated during reductive biotransformation of the azo bond [22]. The azo dyes entering the body by ingestion can be metabolized to aromatic amines by the azoreductases of intestinal microorganisms. If the dyes are nitro, they can be metabolized by the nitroredutases produced by the same microorganisms [29]. Mammalian liver enzymes and other organizations may also catalyze the reductive cleavage of the azo bond and the nitroreduction of the nitro group. In both cases, if N-hydroxylamines are formed, these compounds are capable of causing DNA damage [29, 30].

One of the most difficult tasks confronted by the wastewater treatment plants of textile industries is the removal of the color of these compounds, mainly because dyes and pigments are designed to resist biodegradation, such that they remain in the environment for a long period of time. For example, the half-life of the hydrolyzed dye Reactive Blue 19 is about 46 years at pH 7 and 25°C [31,32].

Carneiro et al. (2010) designed and optimized an accurate and sensitive analytical method for monitoring the dyes C.I. Disperse Blue 373 (DB373), C.I. Disperse Orange 37 (DO37) and C.I. Disperse Violet 93 (DV93) in environmental samples. This investigation showed that DB373, DO37 and DV93 were present in both untreated river water and drinking water, indicating that the effluent treatment (pre-chlorination, flocculation, coagulation and flotation) generally used by drinking water treatment plants, was not entirely effective in removing these dyes. This study was confirmed by the mutagenic activity detected in these wastewaters [33].

In this context, and considering the importance of colored products in present day societies, it is of relevance to optimize the coloring process with the objective of reducing the environmental impact of the textile industry. For this purpose, liposomes could be used to carry several encapsulated dyes, and hence improve the mechanical properties of textile products, resulting in better wash fastness properties and reducing the process temperature, thus economizing energy [34]. Another way is to use ultrasonic energy, studied with the objectives of improving dye productivity and washing fastness, and reducing both energy costs and water consumption [35].

Considering the fact that the textile dyeing process is recognized as one of the most environmentally unfriendly industrial processes, it is of extreme importance to understand the critical points of the dyeing process so as to find alternative, eco-friendly methods.

2. DYEING PROCESS

The dyeing process is one of the key factors in the successful trading of textile products. In addition to the design and beautiful color, the consumer usually looks for some basic product characteristics, such as good fixation with respect to light, perspiration and washing, both initially and after prolonged use. To ensure these properties, the substances that give color to the fiber must show high affinity, uniform color, resistance to fading, and be economically feasible [36].

Modern dyeing technology consists of several steps selected according to the nature of the fiber and properties of the dyes and pigments for use in fabrics, such as chemical structure, classification, commercial availability, fixing properties compatible with the target material to be dyed, economic considerations and many others [36].

Dyeing methods have not changed much with time. Basically water is used to clean, dye and apply auxiliary chemicals to the fabrics, and also to rinse the treated fibers or fabrics [37]. The dyeing process involves three steps: preparation, dyeing and finishing, as follows:

Preparation is the step in which unwanted impurities are removed from the fabrics before dyeing. This can be carried out by cleaning with aqueous alkaline substances and detergents or by applying enzymes. Many fabrics are bleached with hydrogen peroxide or chlorine-containing compounds in order to remove their natural color, and if the fabric is to be sold white and not dyed, optical brightening agents are added [37].

Dyeing is the aqueous application of color to the textile substrates, mainly using synthetic organic dyes and frequently at elevated temperatures and pressures in some of the steps [37,38]. It is important to point out that there is no dye which dyes all existing fibers and no fiber which can be dyed by all known dyes [39]. During this step, the dyes and chemical aids such as surfactants, acids, alkali/bases, electrolytes, carriers, leveling agents, promoting agents, chelating agents, emulsifying oils, softening agents etc [23,38] are applied to the textile to get a uniform depth of color with the color fastness properties suitable for the end use of the fabric [37].This process includes diffusion of the dye into the liquid phase followed by adsorption onto the outer surface of the fibers, and finally diffusion and adsorption on the inner surface of the fibers [40]. Depending on the expected end use of the fabrics, different fastness properties may be required. For instance, swimsuits must not bleed in water and automotive fabrics should not fade after prolonged exposure to sunlight [37]. Different types of dye and chemical additives are used to obtain these properties, which is carried out during the finishing step. Dyeing can also be accomplished by applying pigments (pigments differ from dyes by not showing chemical or physical affinity for the fibers) together with binders (polymers which fix the pigment to the fibers) [39,41].

Finishing involves treatments with chemical compounds aimed at improving the quality of the fabric. Permanent press treatments, water proofing, softening, antistatic protection, soil resistance, stain release and microbial/fungal protection are all examples of fabric treatments applied in the finishing process [37].

Dyeing can be carried out as a continuous or batch process [37]. The most appropriate process to use depends on several factors, such as type of material (fiber, yarn, fabric, fabric construction, garment), generic type of fiber, size of dye batch and quality requirements for the dyed fabric, but batch processes are more commonly used to dye textile materials [10].

In continuous processing, heat and steam are applied to long rolls of fabric as they pass through a series of concentrated chemical solutions. The fabric retains the greater part of the chemicals while rinsing removes most of the preparation chemicals. Each time a fabric is passed through a solution, an amount of water equivalent to the weight of the fabric must be used [37].

In batch processing, sometimes called exhaust dyeing, since the dye is gradually transferred from the dye bath to the material being dyed over a relatively long period of time [10], the dyeing occurs in the presence of dilute chemicals in a closed equipment such as a kier, kettle, beam, jet or beck [37]. Unlike the continuous process, instead of being passed through various baths in a long series of equipment sections, in the batch process the fabric remains in a single piece of equipment, which is alternately filled with water and then drained, at each step of the process. Each time the fabric is exposed to a separate bath, it uses five to ten times its own weight in water [37].

Some batch dyeing machines only operate at temperatures up to 100ºC. However, the system can be pressurized, allowing for the use of temperatures above 100ºC. Cotton, rayon, nylon, wool and some other fibers dye well at temperatures of 100ºC or below. Polyester and some other synthetic fibers dye more easily at temperatures above 100ºC [10].

Since the degree of dye fixation depends on the nature of the fiber, it is important to consider this topic. The fibers used in the textile industry can be divided into two main groups denominated natural fibers and synthetic fibers [36,37]. Natural fibers are derived from the environment (plants or animals), such as wool, cotton, flax, silk, jute, hemp and sisal, most of which are based on cellulose and proteins. On the other hand, synthetic fibers are organic polymers, mostly derived from petroleum sources, for example, polyester, polyamide, rayon, acetate and acrylic [37,39,42]. The two most important textile fibers are cotton, the largest, and polyester [43,44].

Cotton has been used for over 7000 years, and consists of mainly cellulose, natural waxes and proteins. The large number of hydroxyl groups on the cellulose provides a great water absorption capacity [39].

Several aromatic polyesters have been synthesized and studied. Of these, polyethylene terephthalate (PET) and polybutylene terephthalate (PBT) have been produced commercially for more than 50 years. Amongst other uses, PET has been used worldwide for the production of synthetic fibers due to its good physical properties. PET is manufactured from ethylene glycol (EG) and terephthalic acid (TPA) or dimethyl terephthalate (DMT). The polymerization proceeds in two steps: esterification and condensation reactions [45]. The polymer produced after condensation is solidified by jets of cold water and cut into regular granules which frequently have a cubic form. Then, the polymer melt is spun and the yarns are solidified by a stream of cold air [39].

The dye can be fixed to the fiber by several mechanisms, generally in aqueous solution, and may involve primarily four types of interaction: ionic, Van der Waals and hydrogen interactions, and covalent bonds [5].

Ionic interactions result from interactions between oppositely charged ions present in the dyes and fibers, such as those between the positive center of the amino groups and carboxyl groups in the fiber and ionic charges on the dye molecule, and the ionic attraction between dye cations and anionic groups ($-SO_3^-$ and $-CO_2^-$) present in the acrylic fiber polymer molecules. Typical examples of this type of interaction can be found in the dyeing of wool, silk and polyamide [5,36,41].

Van der Waals interactions come from a close approach between the π orbitals of the dye molecule and the fiber, so that the dye molecules are firmly "anchored" to the fiber by an affinity process without forming an actual bond.

Typical examples of this type of interaction are found in the dyeing of wool and polyester with dyes with a high affinity for cellulose [36].

Hydrogen interactions are formed between hydrogen atoms covalently bonded in the dye and free electron pairs of donor atoms in the center of the fiber. This interaction can be found in the dyeing of wool, silk and synthetic fibers such as ethyl cellulose [5,36].

Covalent bonds are formed between dye molecules containing reactive groups (electrophilic groups) and nucleophilic groups on the fiber, for example, the bond between a carbon atom of the reactive dye molecule and an oxygen, nitrogen or sulfur atom of a hydroxy, amino or thiol group present in the textile fiber. This type of bond can be found in the dyeing of cotton fiber [5,36,46].

2.1. Application of liposome-based technology in textile dyeing process

There is increasing interest in the textile industry in the development of eco-friendly textile processing, in which the use of naturally occurring materials such as phospholipids, would become important [47]. Phospholipids are natural surfactants and in the presence of water, they organize themselves so as to reduce unfavorable interactions between their hydrophobic tails and the aqueous solution; their hydrophilic head groups exposed to the aqueous phase forming vesicles. Liposomes or phospholipid vesicles are featured by clearly separate hydrophilic and hydrophobic regions [34,48].

Liposomes were first produced in England in 1961 by Alec D. Bangham, who was studying phospholipids and blood clotting. He found that when phospholipids were added to water, they immediately formed a sphere, because one end of each molecule was water soluble, while the opposite end was water insoluble [49]. From a chemical point of view, the liposome is an amphoteric compound containing both positive and negative charges [34,50].

Liposomes are defined as a structure composed of lipid vesicle bilayers which can encapsulate hydrophobic or hydrophilic compounds in the lipid bilayer or in the aqueous volume, respectively [51]. These structures are usually made up of phosphatidylcholine (PC), which has a hydrophilic part consisting of phosphate and choline groups and a hydrophobic part composed of two hydrocarbon chains of variable length [49, 52,53].

Liposomes are often distinguished according to their number of lamellae and size. Small unilamellar vesicles (SUV), large unilamellar vesicles (LUV) and large multilamellar vesicles (MLV) or multivesicular vesicles (MVV) can be differentiated [49]. The diameters of liposomes vary from a nanometer to a micrometer [34]. Multilamellar liposomes (MLV) usually range from 500 to 10,000 nm. Unilamellar liposomes can be small (SUV) or large (LUV); SUV are usually smaller than 50 nm and LUV are usually larger than 50 nm. Very large liposomes are called giant liposomes (10,000 - 10,00,000 nm). They can be either unilamellar or multilamellar. The liposomes containing encapsulated vesicles are called multi-vesicular and they range from 2,000-40,000 nm. LUVs with an asymmetric distribution of phospholipids in the bilayers are called asymmetric liposomes [54]. The thickness of the membrane (phospholipid bilayer) measures approximately 5 to 6 nm [49].

According to Sivasankar, Katyayani (2011), the preparation of liposomes is based on lipids, and those normally used are:

Natural phospholipids:

- Phosphatidyl choline (PC) - Lecithin
- Phosphatidyl ethanolamine (PE) - Cephalin
- Phosphatidyl serine (PS)
- Phosphatidyl inositol (PI)
- Phosphatidyl glycerol (PG)

Synthetic phospholipids:

For saturated phospholipids:

- Dipalmitoyl phosphatidyl choline (DPPC)
- Distearoyl phosphatidyl choline (DSPC)
- Dipalmitoyl phosphatidyl ethanolamine (DPPE)
- Dipalmitoyl phosphatidyl serine (DPPS)
- Dipalmitoyl phosphatidic acid (DPPA)
- Dipalmitoyl phosphatidyl glycerol (DPPG)

For unsaturated phospholipids:

- Dioleoyl phosphatidyl choline (DOPC)
- Dioleoyl phosphatidyl glycerol (DOPG) [52].

Natural acidic lipids, such as PS, PG, PI, PA (phosphatidic acid) and cardiolipin (CL), are added when anionic liposomes are desired, and cholesterol is often included to stabilize the bilayer. These molecules are derivatives of glycerol with two alkyl groups and one amphoteric group [34].

Phosphatidylcholine is the biological lipid most widely used for producing liposomes. Liposomes based on phosphatidylcholine consist of phosphatidic acid and glycerin, with two alcohol groups esterified by fatty acids and a third group esterified by phosphoric acid, to which the amino alcohol choline is added as a polar group [34,49].

According to Barani, Montazer (2008), normally four different methods can be used for the preparation of liposomes:

1. Dry lipid film;
2. Emulsions;
3. Micelle-forming detergents;
4. Alcohol injection technology [34].

Liposomes have two distinct roles: they can provide an excellent model for biological membranes, and they are being developed as controlled delivery systems for hydrophilic and lipophilic agents [34,53,55]. They are promising candidates for adjuvant and carrier systems for drug delivery, are well-documented, and can be used for the same purpose in textile materials [33].

Encapsulation or liposome technology is applied in numerous fields, such as in pharmaceuticals, cosmetics, foods, detergents, textiles and other applications where it is important to liberate the encapsulated material slowly [34,50]. This new clean technology has already been adopted by some textile industries [52]. In recent years, liposomes have been examined as a way of delivering dyes to textiles in a cost-effective and environmentally sensitive way [56].

Conventional dyeing processes consume a great deal of energy, a significant amount of which is wasted in controlling the process parameters in order to achieve uniform results. With respect to the carrier role of liposomes, they can be used in several textile processes such as textile finishing and dyeing, with several types of dyes and fibers. They are nontoxic, biodegradable, and can encapsulate a wide range of solutes [34]. In addition, the main advantages of liposomes are a clear reduction in dyeing temperature (about 10°C as compared to conventional dyeing), improved quality of the textiles produced, with additional benefits with respect to material weight yield during subsequent spinning, improved smoothness and mechanical properties of the dyed textiles, and a clear reduction in the contamination load of the dye baths

[52,57]. Low temperature gives a more natural feel and improved quality, with lower environmental impact [34].

In recent years, liposomes have been used in the textile industry as a carrier for auxiliary materials (leveling, retarding and wetting agents) in dyeing, mainly for wool dyeing, and for finishing processes [34,55]. One of the most common problems with textile auxiliaries is that they fail to form a complex in the solution bath. This problem can be solved by using liposomes with selected positive or negative charges. Liposomes can be prepared according to the type of process, solute material and fiber structure [34]. Liposomes from phospholipids have been widely used as a dye carrier in the dyeing process, and create eco-friendly textile processes. Due to their structural properties, liposomes can encapsulate hydrophilic dyes (reactive, acid and basic dyes) in the aqueous phase, and hydrophobic dyes (disperse dyes) in the phospholipid bilayers [58]. Liposomes containing a dye are generally large, irregular and unilamellar [50].

According to Barani & Montazer (2008), the application of liposomes in textile processing can be useful when the release of the solute material is important, and improves the final properties of the products. A wetting agent is required in the conventional bleaching bath of cotton fabrics, but this step can be eliminated by using liposomes. The presence of liposomes in the peroxide bleaching bath can improve the mechanical properties of fabrics and their brightness. Liposomes contain particles of oxidant present in the bleaching solution that represent an unusual reservoir, and release the bleaching agent gradually into the bleaching bath. Moreover, the encapsulation of catalysts used for the decomposition of hydrogen peroxide radicals can be another factor in retarding the rate of decomposition. In this way, liposomes act as a stabilizing agent in the bleaching bath [34].

The role of auxiliary products is very important in textile dyeing with disperse dyes [53]. These compounds show extremely low solubility in water and dispersing agents are needed to maintain a fine, stable dispersion throughout the whole dyeing process at the different temperatures. Martí et al. (2007) analyzed the usefulness of commercial textile liposomes as dispersing agents, and observed that liposomes could be considered as suitable dispersing auxiliaries for polyester dyeing at high temperatures, considering their capacity to stabilize dye dispersions and achieve a suitable dye exhaustion level, with the added value of their environmentally friendly nature [53]. Liposomes clearly improve the dispersion efficiency as compared to conventional dispersing agents [34].

Additionally, liposomes for textile use show a similar price to that of synthetic surfactants used in the dyeing of polyester with disperse dyes. However, the new technology is more environmentally friendly, and hence the reduction in the environmental problem can lead to economic advantages [53]. In addition, liposome preparations tend not to foam. This is an advantage that distinguishes liposomes from other textile auxiliaries [34].

According to Martí et al. (2010), the dyeing of wool and wool blends with the aid of liposomes has demonstrated better quality, energy saving and a reduction in the environmental impact and also the temperature could be reduced, resulting in less fiber damage. Moreover, dye bath exhaustion was shown to be over 90% at the lower temperature (80°C) used, resulting in significant savings in energy costs [55]. The impact of the dyeing process on the environment was also considerably lower, the COD being reduced by about 1000 units [53].

Therefore, liposome-based technology is an alternative, eco-friendly method, which could reduce the environmental impact, offering technical and economic advantages for the textile industry.

2.2. Effect of ultrasonic energy on the dyeing process

Ultrasound-assisted textile dyeing was first reported by Sokolov and Tumansky in 1941[59]. The basic idea of this technology is that ultrasound can enhance mass transfer by reducing the stagnant cores in the yarns. The improvements observed are generally attributed to cavitation phenomena and to other resulting physical effects such as dye dispersion (breaking up of aggregates with high relative molecular mass), degassing (expulsion of dissolved or entrapped air from the fiber capillaries), strong agitation of the liquid (reduction in thickness of the fiber-liquid boundary layer), and swelling (enhancement of dye diffusion rate inside the fiber) [59,60].

According to Vankar & Shanker (2008), ultrasound allows for process acceleration, obtaining the same or better results than existing techniques, but under less extreme conditions, i.e., lower temperatures and lower concentrations of the chemicals used. Wet textile processes assisted by ultrasound are of great interest to the textile industry for this reason [61], and Khatri et al. (2011) showed that the dyeing of polyester fiber using ultrasonic energy resulted in an increased dye uptake and enhanced dyeing rate [35].

Due to the revolution in environmental protection, the use of ultrasonic energy as a renewable source of energy in textile dyeing has been increased, due to the variety of advantages associated with it. On the other hand, there is a growing demand for natural, eco-friendly dyeing for the health sensitive application to textile garments as an alternative to harmful synthetic dyes, which poses a need for suitable effective dyeing methodologies [62].

Ultrasonic energy can clean or homogenize materials, accelerating both physical and chemical reactions, and these qualities can be used to improve textile processing methods. Environmental concern has been focused on textile processing methods for quite some time, and the use of ultrasonic energy has been widely studied in terms of improving washing fastness. The textile dyeing industry has long been struggling to cope with high energy costs, rapid technological changes and the need for a faster delivery time, and the effective management of ultrasonic energy could reduce energy costs and improve productivity [35]. Ultrasonic waves are vibrations with frequencies above 17 kHz, out of the audible range for humans, requiring a medium with elastic properties for propagation. The formation and collapse of the bubbles formed by ultrasonic waves (known as cavitation) is generally considered to be responsible for most of the physical and chemical effects of ultrasound in solid/liquid or liquid/liquid systems [63]. Cavitation is the formation of gas-filled microbubbles or cavities in a liquid, their growth, and under proper conditions, their implosive collapse [59].

It has been reported that ultrasonic energy can be applied successfully to wet textile processes, for example laundering, desizing, scouring, bleaching, mercerization of cotton fabrics, enzymatic treatment, dyeing and leather processing, together with the decoloration/mineralization of textile dyes in waste water [60].

In addition, ultrasonic irradiation shows promise, and has the potential, for use in environmental remediation, due to the formation of highly concentrated oxidizing species such as hydroxyl radicals ($HO\bullet$), hydrogen radicals ($H\bullet$), hydroperoxyl radicals ($HO_2\bullet$) and H_2O_2, and localized high temperatures and pressures [59]. Therefore, the use of ultrasonic energy could indeed reduce the environmental impact caused by the textile industry.

3. FINISHING AND WASTE WATER

The contamination of natural waters has become one of the biggest problems in modern society, and the economical use of this natural resource in production processes has gained special attention, since in predictions for the coming years, the amount of water required per capita is of concern. This environmental problem is related not only to its waste through misuse, but also to the release of industrial and domestic effluents [64].

Of the industries with high-polluting power, the textile dyeing industry, responsible for dyeing various types of fiber, stands out. Independent of the characteristics of the dyes chosen, the final operation of all dyeing process involves washing in baths to remove excesses of the original or hydrolyzed dyes not fixed to the fiber in the previous steps [36]. In these baths, as previously mentioned, it is estimated that approximately 10-50% of the dyes used in the dyeing process are lost, and end up in the effluent [17,21,22], contaminating the environment with about one million tons of these compounds [65]. The dyes end up in the water bodies due mainly to the use of the activated sludge treatment in the effluent treatment plants, which has been shown to be ineffective in removing the toxicity and coloring of some types of dye [33,60,66,67]. Moreover, the reduction of azo dyes by sodium hydrosulfite and the successive chlorination steps with hypochlorous acid, can form 2-benzotriazoles fenilbenzotriazol (PBTA) derivatives and highly mutagenic aromatic amines, often more mutagenic than the original dye [68]. In an aquatic environment, this dye reduction can occur in two phases: 1) The application of reducing agents to the newly-dyed fibers to remove the excess unbound dye, which could lead to "bleeding" of the fabrics during washing, and 2) The use of reducing agents in the bleaching process, in order to make the effluent colorless and conform with the legislation. This reduced colorless effluent containing dyes is sent to the municipal sewage treatment plant, where they chlorinate the effluents before releasing them into water bodies where they may generate PBTAs. Several different PBTAs are already described in the literature, and their chemical structures vary depending on the dyes that originated them [63,69].

So the release of improperly treated textile effluents into the environment can become an important source of problems for human and environmental health. The major source of dye loss corresponds to the incomplete fixation of the dyes during the textile fiber dyeing step [36].

In addition to the problem caused by the loss of dye during the dyeing process, within the context of environmental pollution, the textile industry is also focused due to the large volumes of water used by its industrial park, consequently generating large volumes of effluent [64]. It has been calculated that approximately 200 liters of water are needed for each kilogram of cotton produced [70]. These effluents are complex mixtures of many pollutants, ranging from original colors lost during the dyeing process, to associated pesticides and heavy metals [71], and when not properly treated, can cause serious contamination of the water sources [64]. So the materials that end up in the water bodies are effluents containing a high organic load and biochemical oxygen demand, low dissolved oxygen concentrations, strong color and low biodegradability. In addition to visual pollution, the pollution of water bodies with these compounds causes changes in the biological cycles of the aquatic biota, particularly affecting the photosynthesis and oxygenation processes of the water body, for example by hindering the passage of sunlight through the water [72].

Moreover, studies have shown that some classes of dye, especially azo dyes and their by-products, may be carcinogenic and / or mutagenic [27,33,67,73-

77], endangering human health, since the wastewater treatment systems and water treatment plants (WTP) are ineffective in removing the color and the mutagenic properties of some dyes [78,79]. The difficulty in removing them from the environment can be attributed to the high stability of these compounds, since they are designed to resist biodegradation to meet the demands of the consumer market with respect to durability of the colors in the fibers, consequently implying that they also remain in the environment for a long time [32].

With respect to the legislation, there is no consensus amongst the different countries concerning effluent discharge, and there is no official document listing the different effluent limit values applied in different countries. Many federal countries, such as the United States of America, Canada and Australia have national environmental legislation, which, as in Europe, establishes the limits that must be complied with. Some countries, such as Thailand, have copied the American system, whereas others, such as Turkey or Morocco, have copied the European model. In some countries, for example India, Pakistan and Malaysia, the emission limits are recommended, but are not mandatory [80]. With respect to the color, in some countries such as France, Austria and Italy, there are limits for the color of the effluent, but since they use different units, a comparison is impossible. The oldest unit is the Hazen, in use since the beginning of the 20th century, but in France, the current unit is (mg L^{-1}Pt–Co). The coloration values are determined by a comparative analysis with model solutions prepared according to defined procedures [80].

Based on all the problems cited above regarding the discharge of effluents into the environment, it is obvious there is a need to find alternative treatments that are effective in removing dyes from effluents.

4. WASTE WATER TREATMENT

These days, environmental pollution can undoubtedly be regarded as one of the main problems in developed and developing countries. This is due, not just to one, but to a number of factors, such as the misuse of natural resources, inefficient legislation and a lack of environmental awareness. Fortunately, in recent years there has been a trend for change and a series of scientific studies are being used as an important tool in the development of new treatment technologies and even in the implementation of processes and environmentally friendly actions [64,81-84].

Every industrial process is characterized by the use of inputs (raw materials, water, energy, etc.) that undergo transformation giving rise to products, byproducts and waste. The wastes produced at all stages of the various types of human activity, both in terms of composition and volume, vary according to the consumption practices and production methods. The main concerns are focused on the impact these can have on human health and the environment. Hazardous waste, produced mainly by industry, is particularly worrying, because when incorrectly managed, it becomes a serious threat to the environment and therefore to human health. Thus the study of new alternatives for the treatment of different types of industrial effluent continues to be a challenge to combat anthropogenic contamination.

Amongst that of several other industries, the textile sector waste has received considerable attention in recent years, since it can generate large volumes of effluents that, if not correctly treated before being disposed into water resources, can be a problem, as previously mentioned. Effluents from the textile industry are extremely complex, since they contain a large variety of dyes, additives and derivatives that change seasonally, increasing the challenge to find effective, feasible treatments. Currently, the processes developed and available for these industries are based on methods that were designed for other waste, and have limitations when applied to textile effluents. As a consequence, these industries produce colored wastewater with a high organic load, which can contribute enormously to the environmental pollution of surface water and treatment plants if not properly treated before disposal into the water resources [85]. The ingestion of water contaminated with textile dyes can cause serious damage to the health of humans and of other living organisms, due to the toxicity, highlighting mutagenicity of its components [86,87]. Therefore treatments that are more efficient and economical than those currently available are required.

There are several techniques for the treatment of effluents, such as incineration, biological treatment, absorption onto solid matrices, etc. However, these techniques have their drawbacks, such as the formation of dioxins and furans, caused by incomplete combustion during incineration; long periods for biological treatment to have an effect, as also the adsorptive process, that is based on the phase transfer of contaminants without actually destroying them [88,89]. The problem is further aggravated in the textile industry effluents, due to the complexity of their make-up. Thus it can be seen that processes are being used that are not entirely appropriate for the treatment of textile effluents, thereby creating a major challenge for the industry and laundries that need to adapt to current regulations for the control of the color of effluents with a high organic load.

The use of filtration membranes and/or separation [90] and biological methods [91], in addition to incineration processes involving adsorption onto solid matrices, has also being adopted by the textile industry and is receiving considerable attention. However, all these processes only involve phase transfer, generating large amounts of sludge deposited at the end of the tanks and low efficiency in color removal and reduction of the organic load.

According to this scenario, many studies have been carried out with the aim of developing new technologies capable of minimizing the volume and toxicity of industrial effluents. Unfortunately, the applicability of these types of system is subject to the development of modified procedures and the establishment of effluent recycling systems, activities that imply evolutionary technologies and which are not yet universally available. Thus the study of new alternatives for the treatment of many industrial effluents currently produced is still one of the main weapons to combat the phenomenon of anthropogenic contamination.

Due to their considerable danger, several authors have attempted to find new forms of treatment to reduce the serious environmental and toxicological risks caused by various organic compounds. Amongst the many reported cases are those based on the use of specific microorganisms, and degradation using advanced oxidation processes (AOP) such as Fenton, photo-Fenton and heterogeneous photocatalysis, which are highlighted below.

The use of microorganisms cultivated specifically for the degradation of pollutants to increase the yield of degradation, has been reported by some authors. For example, Flores et al. (1997) examined the behavior of 25 N-substituted aromatic compounds such as organic compounds,azo dyes and nitro, using the methanogenic bacteria acetoclastic, and found that under anaerobic conditions it was easy to mineralize various of the compounds evaluated with a good yield, especially the nitroaromatic and azo dyes [91].

Bornick et al. (2001) evaluated the use of aerobic microorganisms to degrade aromatic amines present in sediments of the river Elbe in Germany. The results obtained showed that it was possible to predict the qualitative degradation of the aromatic amines using degradation constants [92].

Using a structural design, Wang et al.(2007) isolated a bacterium capable of promoting the degradation of the compounds pentyl amine and aniline present in water oil extraction in China. Under conditions of neutral pH and complete aeration of 6 mg O_2 / l at a temperature of 30 ° C, they obtained degradation yields of 82% and 78%, respectively, for pentyl amine and aniline [93].

However, in general, although the use of microorganisms in the treatment of industrial and laboratory wastes containing aromatic amines deserves attention, mainly due to the low investment and maintenance costs, the results are far from ideal, due to the low biodegradation yields, long treatment times, and the generation of sludge deposited at the bottom of the treatment ponds [94].

Of the studies carried out using a source of hydroxyl radicals with oxidizing agents, the photo Fenton system developed by Fukushima et al., 2000 to promote the degradation of aniline stands out. This method has shown promise for the mineralization of aromatic amines, obtaining a reduction of approximately 85%. However, high performance liquid chromatography (HPLC) identified a number of intermediate species formed during the degradation of the aniline, such as p-aminophenol, p-hydroquinone, maleic and fumaric acids and NH_4^+ [95].

Studies involving heterogeneous photocatalysis also deserve attention: for example Pramauro et al., (1995) promoted the degradation of various aniline derivatives using TiO_2 particles suspended in a solution. Under optimal conditions, the method developed showed rapid mineralization of the aromatic amines examined in less than 1 hour of analysis. By-products generated during the degradation of these compounds were identified at the start of the reaction, but none were identified at the end of the reaction. The use of solar radiation was also evaluated, but was shown to be less efficient than artificial radiation [96].

Augugliaro et al.(2000), confirmed that heterogeneous photocatalysis using TiO_2 as a semiconductor may be a suitable method for the complete photodegradation of aniline, 4-ethylaniline and 4-chloroaniline in an aqueous medium. The kinetic parameters for the Langmuir-Hinshelwood model were used to describe the importance of the adsorption results, which proved to be independent of the pH of the solution and of the type of substituent on the aromatic ring of the amine [97].

Canle et al. (2005) studied the behavior of the adsorption of aniline and dimethyl aniline onto three species of TiO_2 (P25, anatase and rutile) and established that the greatest adsorption occurred onto TiO_2 P25. The effect of pH on the degradation provided by these compounds was also evaluated, and lower mineralization percentages were observed in acidic media. This type of behavior can be attributed to the positive charges on both the aniline and the semiconductor, providing electrostatic repulsion between the two species. Thus, an alkaline medium has been recommended as the most appropriate one to promote the mineralization of aromatic amines using a TiO_2 P25 type semiconductor [98].

Chu et al. (2007) observed the effects of pH variation and the addition of hydrogen peroxide on the degradation of 2-chloroaniline using TiO_2 as a semiconductor, with and without the application of UV radiation. The results showed that the addition of low concentrations of H_2O_2 to the UV/TiO_2system provided a significant increase in degradation of the aromatic amine. The addition of an excess of H_2O_2 promoted no increase in degradation, as expected; to the contrary, a reduction in the reaction rate was observed. The variation in pH was evaluated in both systems, and the condition leading to the highest percentages of mineralization was obtained in an alkaline medium using H_2O_2/UV/TiO_2 [99].

Low et al. (1991) monitored the inorganic products resulting from the degradation of several organic nitrogenated, sulfured and halogenated compounds. Degradation was carried out using TiO_2 as the semiconductor and artificially illuminated UV radiation. Ammonium ions were found to be present in higher concentrations than nitrate ions, which can be explained by the fact that compounds having a nitrogen element in their structure pass through a complex degradation step where the generation of ammonium ions is more favorable than the generation of nitrate ions. In turn, compounds with nitro groups in their structures, showed higher concentrations of nitrate ions. Under ideal conditions, all the elements were converted into their respective inorganic forms. Organic carbon was converted to CO_2, the halogens to their corresponding halide, sulfur compounds to sulfate, the phosphate to phosphorus and nitrogen to ammonium and nitrate [100]. However, all these studies used a photocatalytic titanium suspension, requiring a subsequent step to remove the semiconductor.

With a view to these problems, a new technique that has been studied recently, with significant success, is the oxidation of organic matter via the generation of hydroxyl radicals (OH•). This kind of effluent treatment has been highlighted for its destructive character with respect to the organic matter. The advanced oxidation processes (AOP) are processes based on the generation of highly oxidizing species, the hydroxyl radical (OH•), which can oxidize the organic contaminants present in water, air or soil. These radicals have a high oxidation potential (E * = +2.72 V vs. the normal hydrogen electrode, NHE), which results in high reactivity with organic pollutants. This may initiate different types of reaction with different functional groups in organic compounds, forming unstable organic radicals which are then easily oxidized to CO_2, H_2O and inorganic acids, derived from this heteroatom. Many techniques have being developed using this principle as the method of treatment, and have shown potential success in organic residue mineralization processes [92,96, 101-103].

The association of the electrochemical properties with the photocatalytic properties of a semiconductor, have allowed for the development of a promising system, representing a very efficient technique as an AOP method that has received much attention in recent years, assigned as a photoelectrochemical process. The process gained notoriety for the possibility of forming hydroxyl radicals via the oxidation of water. The technique is based on the action of ultraviolet light (hu) on a semiconductor capable of generating charges and e^-/ H^+, whose separation is facilitated by applying a positive potential (E_{APP}), greater than the potential of a flat band photocatalytic material. The generation of a potential gradient in the photoactivated semiconductor directs the electrons to an auxiliary electrode (cathode), delaying recombination between the holes (h^+) generated in the valence band

(VB), and making them available for oxidation processes and the generation of hydroxyl radicals of interest [104].

The results of this process were very promising because of the relatively short treatment time but with great efficiency, both in the removal of color and in the reduction of the organic load. However, the limitations of this technique are related mainly to the choice of the ideal catalyst for promoting the generation of these oxidizing species. Catalysts that promote the generation of radicals absorbing radiation in the visible spectral region are the most desirable for this type of reaction, due to the large percentage emitted in the solar spectrum (approximately 45%) [104].

Thus, the development of an ideal process that promotes color removal and a reduction in the organic load of wastewater from the textile industry with great efficiency is a major challenge in all fields of science, since the synthesis of the best catalyst to take advantage of solar radiation, thus reducing the operating costs, and at the same time solve the problems involved in the hydrodynamics of the reactors, is of importance in the development of the treatment.

The expectations for developing an effective method for the treatment of these wastes are quite promising, but require continuous optimization and knowledge of new aspects. These include better fixation of the dyes to the fibers, process with less water consumption, less hazardous dyes with respect to human health and methods capable of identifying these compounds with more efficacy and rapidity and assays to identify any potential carcinogenic and / or mutagenic properties in the dyes and their derivatives; genetic improvements to produce more efficient culture mediums and resistant biological treatments, leading to a reduction in the generation of sludge; the synthesis of materials that catalyze reactions in the visible spectral regions, leading to a more economic photoeletrochemical method, and also new engineering advances for the construction of more effective reactors, which can take advantage of all these developments in an integrated system, extending the performance of a process more appropriate for the treatment of such a complex effluent.

5. OPTIMIZATION OF THE DYEING PROCESSES TO REDUCE THE ENVIRONMENTAL IMPACT OF THE TEXTILE INDUSTRY

The search and development of new methods to promote the treatment of effluents from the textile industry with a maximum of efficiency of the process of decolorization and / or removal of these compounds present in the medium can trigger further damage human health and the environment is fundamental importance. The understanding of the composition of waste generated is extremely significant to develop these methods of treatment due to the high complexity by virtue of huge number of compounds which are added at different stages of the dyeing fabrics.

Environmental problems with used dye baths are related to the wide variety of different components added to the dye bath, often in relatively high concentrations. In the future, many of textile factories will face the requirement of reusing a significant part of all incoming freshwater because traditionally used methods are insufficient for obtaining the required water quality.

However, due to dwindling supply and increasing demand of water in the textile industries, a better alternative is to attempt to further elevate the water quality of wastewater effluent from a secondary wastewater treatment plant to a higher standard for reuse. Thus far very little attention has been paid to this aspect [105].

Therefore, thc investment in the search for methodologies to more effective treatment of these effluents can be much smaller than that spent in tertiary treatment to remove these products in low level of concentrations and in the presence of much other interference. This requires action that the cost / benefit are reviewed and the development of new techniques for wastewater treatment capable of effective removal of these dyes is intensified and made economically viable [105,106].

An alternative to minimize the problems related to the treatment of textile effluents would be the development of more effective dye that can be fixed fiber with higher efficiency decreasing losses on tailings waters and reducing the amount of dye required in the dyeing process, reducing certainly improve the cost and quality of the effluent.

6. CONCLUSION

It was concluded that the synthetic textile dyes represent a large group of organic compounds that could have undesirable effects on the environment, and in addition, some of them can pose risks to humans. The increasing complexity and difficulty in treating textile wastes has led to a constant search for new methods that are effective and economically viable. However, up to the present moment, no efficient method capable of removing both the color and the toxic properties of the dyes released into the environment has been found.

ACKNOWLEDGMENTS

This work was supported by the Faculty of Pharmaceutical Sciences at Ribeirão Preto - University of São Paulo, Brazil, and by FAPESP, CAPES and CNPq.

REFERENCES

1. Kirk-Othmer. Encyclopedia of Chemical Technology, v. 7, 5th Edition. Wiley-Interscience; 2004.

2. Bafana A, Devi SS, Chakrabarti T. Azo dyes: past, present and the future. Environmental Reviews 2011; 19 350–370.

3. Zollinger, H. Synthesis, Properties of Organic Dyes and Pigments. In: Color Chemistry. New York, USA: VCH Publishers; 1987. p. 92-102.

4. Carneiro PA, Nogueira, RFP, Zanoni, MVB. Homogeneous photodegradation of C.I. Reactive Blue 4 using a photo-Fenton process under artificial and solar irradiation. Dyes and Pigments 2007; 74 127-132.

5. Christie R. Colour Chemistry. Cambridge, United Kingdom: The Royal Society of Chemistry; 2001.

6. Dos Santos AB, Cervantes FJ, van Lier JB. Review paper on current technologies for decolourisation of textile wastewaters: Perspectives for anaerobic biotechnology. Bioresource Technology 2007; 98 (12) 2369-2385.

7. Arun Prasad AS, Bhaskara Rao KV. Physico chemical characterization of textile effluent and screening for dye decolorizing bactéria. Global Journal of Biotechnology and Biochemistry 2010; 5(2) 80-86.

8. Robinson T, McMullan G, Marchant R, Nigam P. Remediation of dyes in textile effluent: a critical review on current treatment technologies with a proposed alternative. Bioresource Technology 2001; 77 (12) 247-255.

9. Ogugbue CJ, Sawidis T. Bioremediation and Detoxification of Synthetic Wastewater Containing Triarylmethane Dyes by Aeromonas hydrophila Isolated from Industrial Effluent. Biotechnology Research International 2011; DOI 10.4061/2011/967925.

10. Perkins WS. A Review of Textile Dyeing Processes. American Association of Textile Chemists and Colorists 1991; 23 (8): 23–

27.http://www.revistavirtualpro.com/files/TIE08_200704.pdf (accessed 12 May 2012)

11. Couto SR. Dye removal by immobilised fungi. Biotechnology Advances 2009; 27(3) 227-235.

12. O'Neill C, Hawkes FR, Hawkes DL, Lourenço ND, Pinheiro HM, Delée W. Colour in textile effluents – sources, measurement, discharge consents and simulation: a review. Journal of Chemical Technology and Biotechnology 1999; 74 (11) 1009-1018.

13. Forgacs E, Cserháti T, Oros G. Removal of synthetic dyes from wastewaters: a review. Environment International 2004; 30 (7) 953-971

14. Przystaś W, Zabłocka-Godlewska E, Grabińska-Sota E. Biological Removal of Azo and Triphenylmethane Dyes and Toxicity of Process By-Products. Water Air Soil Pollut 2012; 223 (4) 1581-1592.

15. Sen S, Demirer GN. Anaerobic treatment of real textile wastewater with a fluidized bed reactor. Water Research 2003; 37 (8) 1868-1878.

16. Dos Santos, A B. Reductive Decolourisation of Dyes by Thermophilic Anaerobic Granular Sludge. PhD-Thesis. Wageningen University, The Netherlands, 2005.

17. Ben Mansour H, Houas I, Montassar F, Ghedira K, Barillier D, Mosrati R, Chekir-Ghedira L. Alteration of in vitro and acute in vivo toxicity of textile dyeing wastewater after chemical and biological remediation. Environmental science and pollution research international 2012; DOI 10.1007/s11356-012-0802-7.

18. Talarposhti AM, Donnelly T, Anderson GK. Colour removal from a simulated dye wastewater using a two-phase anaerobic packed bed reactor. Water Research 2001; 35 (2) 425-432.

19. Ibrahim MB, Poonam N, Datel S, Roger M. Microbial decolorization of textile dye-containing effluents: a review, Bioresource Technology 1996; 58(3) 217-227.

20. Wijetunga S, Li XF, Jian C. Effect of organic load on decolourization of textile wastewater containing acid dyes in upflow anaerobic sludge blanket reactor. Journal of Hazardous Materials 2010; 177 (1-3) 792-798.

21. Vaidya AA, Datye KV. Environmental pollution during chemical processing of synthetic fibers. Colourage 1982; 14 3-10.

22. Rajaguru P, Fairbairn LJ, Ashby J, Willington MA, Turner S, Woolford LA, Chinnasamy N, Rafferty JA. Genotoxicity studies on the azo dye Direct Red 2 using the in vivo mouse bone marrow micronucleus test. Mutation Research 1999; 444(1) 175-180.

23. Hubbe MA, Beck KR, O'Neal WG, Sharma YC. Cellulosic substrates for removal of pollutants from aqueous systems: a review. 2. Dyes. Dye biosorption: Review. BioResources 2012; 7 (2), 2592-2687.

24. Carliell CM, Barclay SJ, Shaw C, Wheatley AD, Buckley C A. The effect of salts used in textile dyeing on microbial decolourisation of a reactive azo dye. Environmental Technology 1998; 19 (11) 1133-1137.

25. Seesuriyachan P, Takenaka S, Kuntiya A, Klayraung S, Murakami S, Aoki K. Metabolism of azo dyes by Lactobacillus casei TISTR 1500 and effects of various factors on decolorization. Water Research 2007; 41(5) 985-992.

26. Ben Mansour H, Corroler D, Barillier D, Ghedira K, Chekir L, Mosrati R. Evaluation of genotoxicity and pro-oxidant effect of the azo dyes: Acids yellow 17, violet 7 and orange 52, and of their degradation products by Pseudomonas putida mt-2. Food and Chemical Toxicology 2007; 45 (9) 1670-1677.

27. Chung KT, Cerniglia CE. Mutagenicity of azo dyes: Structure-activity relationships. Mutation Research, 1992; 277 (3) 201-220.

28. Pinheiro HM, Touraud E, Thomas O. Aromatic amines from azo dye reduction: status review with emphasis on direct UV spectrophotometric detection in textile industry wastewaters. Dyes and Pigments 2004; 61 (2) 121-139.

29. Umbuzeiro GA, Freeman H, Warren SH, Kummrow F, Claxton LD. Mutagenicity evaluation of the commercial product CI Disperse Blue 291 using different protocols of the Salmonella assay. Food and Chemical Toxicology 2005; 43 (1) 49-56.

30. Arlt VM, Glatt H, Muckel E, Pabel U, Sorg BL, Schmeiser HH, Phillips DH. Metabolic activation of the environmental contaminant 3-nitrobenzanthrone by human acetyltransferases and sulfotransferase. Carcinogenesis 2002; 23 (11) 1937-1945.

31. Hao OJ, Kim H, Chiang PC, 2000. Decolorization of wastewater. Critical Reviews in Environmental Science and Technology 2000; 30 (4) 449-505.

32. Firmino PIM, Silva MER, Cervantes FJ, Santos AB. Colour removal of dyes from synthetic and real textile wastewaters in one- and two-stage anaerobic systems. Bioresource Technology 2010; 101(20) 7773-7779.

33. Carneiro PA, Umbuzeiro GA, Oliveira DP, Zanoni MVB. Assessment of water contamination caused by a mutagenic textile effluent/dyehouse effluent bearing disperse dyes. Journal of Hazardous Materials 2010; 174 (1-3) 694-699.

34. Barani H, Montazer M. A Review on Applications of Liposomes in Textile Processing. Journal of Liposome Research 2008; 18 (3) 249-262.

35. Khatri Z, Memon MH, Khatri A, Tanwari A. Cold Pad-Batch dyeing method for cotton fabric dyeing with reactive dyes using ultrasonic energy. Ultrasonics Sonochemistry 2011 18 (6) 1301-1307.

36. Guaratini CCI, Zanoni MVB. Textile dyes. Química Nova 2000;23(1): 71-78. http://www.scielo.br/pdf/qn/v23n1/2146.pdf (accessed 01 May 2012).

37. Moore SB, Ausley LW. Systems thinking and green chemistry in the textile industry: concepts, technologies and benefits. Journal of Cleaner Production 2004;12 585–601. DOI: 10.1016/S0959-6526(03)00058-1

38. Reddy SS, Kotaiah B, Reddy NSP. Color pollution control in textile dyeing industry effluents using tannery sludge derived activated carbon. Bulletin of the Chemical Society of Ethiopia 2008;22 (3): 369-378. http://www.ajol.info/index.php/bcse/article/viewFile/61211/49389 (accessed 16 May 2012).

39. Alcantara MR, Daltin D. A química do processamento têxtil. Química Nova 1996;19(3): 320-330 http://www.quimicanova.sbq.org.br/qn/qnol/1996/vol19n3/v19_n3_17.pdf (accessed 19 April 2012).

40. Vassileva V, Valcheva E, Zheleva Z. The kinetic model of reactive dye fixation on cotton fibers. Journal of the University of Chemical Technology and Metallurgy 2008;43 (3): 323-326. http://www.uctm.edu/j2008-3/8_V_Vasileva_323-326.pdf (accessed 10 May 2012).

41. Zollinger H. Color Chemistry. Syntheses,properties and applications of organic dyes and pigments. Weinheim: VCH Publishers; 2003.

42. Candlin J. Polymers. Dyestuffs in the Chemical Industry. London: Chapman & Hall; 1994.

43. Gregory P. Dyestuffs. In: Heaton C.A. (ed.) The Chemical Industry. London: Chapman & Hall; 1994. p143-188.

44. Gregory P. 2009. Dyes and Dye Intermediates. In: Kirk-Othmer (ed.) Encyclopedia of Chemical Technology. New Jersey: John Wiley & Sons; 2009. p1–66. DOI: 10.1002/0471238961.0425051907180507.a01.pub2

45. Pang K, Kotek R, Tonelli A. Review of conventional and novel polymerization processes for polyesters. Progress in Polymer Science 2006;31(11) 1009–1037. DOI: 10.1016/j.progpolymsci.2006.08.008

46. Hunger K., editor. Industrial Dyes: Chemistry, Properties, Applications. Weinheim: Wiley-VCH Verlag GmbH & Co. KGaA; 2004. DOI: 10.1002/3527602011

47. Baptista ALF, Coutinho PJG,: Real Oliveira MECD, Rocha Gomes JIN. Effect of surfactants in soybean Lecithin liposomes studied by energy transfer between NBD-PE and N-Rh-PE. Journal of Liposome Research 2000; 10 (4) 419-429.

48. Lesoin L, Crampon C, Boutin O, Badens E. Preparation of liposomes using the supercritical anti-solvent (SAS) process and comparison with a conventional method. The Journal of Supercritical Fluids 2011; 57 162-174

49. Sivasankar, M, Katyayani, T. Liposomes – The future of formulations. International Journal of Research in Pharmacy and Chemistry 2011; 1(2) 259-267.

50. El-Zawahry MM, El-Shami S, El-Mallah S.H. Optimizing a wool dyeing process with reactive dye by liposome microencapsulation. Dyes and Pigments 2007; 74 684-691.

51. Martí M, de la Maza A, Parra JL, Coderch L. Liposome as dispersing agent into disperse dye formulation. Textile Research Journal 2011; 81(4) 379-387.

52. Montazer M., Validi M., Toliyat T. Influence of Temperature on Stability of Multilamellar Liposomes in Wool Dyeing. Journal of Liposome Research, 2006; 16:81-89.

53. Martí M, Coderch L, de la Maza A, Parra JL. Liposomes of phosphatidylcholine: a biological natural surfactant as a dispersing agent. Color Technology 2007; 123 237-241.

54. Riaz M. Liposomes preparation methods. Pakistan Journal of Pharmaceutical Sciences, 1996; 19(1) 65-77.

55. Martí M, de la Maza A, Parra JL, Coderch L. Liposome as dispersing agent into disperse dye formulation. Textile Research Journal 2010; 81(4) 379–387.

56. Nelson, G. Application of microencapsulation in textile. International Journal of Pharmaceutics 2002; 242 55-62.

57. Martí M, de la Maza A, Parra JL, Coderch L. Dyeing wool at low temperatures: new method using liposomes. Textile Research Journal 2001; 71(8) 678-682.

58. El-Zawahry MM, El-Mallahb MH, El-Shamib S. An innovative study on dyeing silk fabrics by modified phospholipid liposomes. Coloration Technology 2009; 125 164-171.

59. Vajnhandl S, Le Marechal AM. Ultrasound in textile dyeing and the decolouration/mineralization of textile dyes. Dyes and Pigments 2005; 65 89-101.

60. Ferrero F., Periolatto M. Ultrasound for low temperature dyeing of wool with acid dye. Ultrasonics Sonochemistry 2012; 19 601-606.

61. Vankar PS, Shanker R. Ecofriendly ultrasonic natural dyeing of cotton fabric with enzyme pretreatments. Desalination 2008; 230 62-69.

62. Mansour HF, Heffernan S. Environmental aspects on dyeing silk fabric with sticta coronate lichen using ultrasonic energy and mild mordants. Clean Technologies Environmental Policy 2011; 13 207-213.

63. Abou-Okeil A, El-Shafie A, El Zawahry MM. Ecofriendly laccase–hydrogen peroxide/ultrasound-assisted bleaching of linen fabrics and its influence on dyeing efficiency. Ultrasonics Sonochemistry 2010; 17 383-390.

64. Kunz A, Zamora PP, Moraes SG, Durán N. New tendencies on the textile effluent treatment. Quimica Nova 2002;25(1): 78-82. http://www.scielo.br/scielo.php?script=sci_arttext&pid=S0100-40422002000100014 (accessed 01 May 2012).

65. Associação Brasileira de Química. ABQ. Processos oxidativos avançados e tratamento de corantes sintéticos. Anais da ABQ. http://www.unb.br/resqui/abq2004-1.pdf (accessed 18 May 2012).

66. Umbuzeiro GA, Roubicek DA, Rech CM, Sato MIZ, CLAXTON LD. Investigating the sources of the mutagenic activity found in a river using the Salmonella assay and different water extraction procedures. Chemosphere 2004; 54(11) 1589-1597. DOI: 10.1016/j.chemosphere.2003.09.009

67. Umbuzeiro GA, Freeman HS, Warren SH, Oliveira DP, Terao Y, Watanabe T, Claxton LD. The contribution of azo dyes to the mutagenic activity of the Cristais river. Chemosphere 2005; 60 (1) 55-64. DOI:10.1016/j.chemosphere.2004.11.100

68. Shiozawa T, Suyama K, Nakano K, Nukaya H, Sawanishi H, Oguri A, Wakabayashi K, Terao Y. Mutagenic activity of 2 phenylbenzotriazole derivatives related to a mutagen, PBTA-1, in river water. Mutation Research 1999; 442 (2) 105-111.

69. Oliveira DP. Corantes como importante classe de contaminantes ambientais – um estudo de caso. PhD thesis. Universidade de São Paulo; 2005.

70. Carneiro PA, Osugi ME, Sene JJ, Anderson MA, Zanoni MVB. Evaluation of color removal and degradation of a reactive textile azo dye on nanoporous TiO thin-film electrodes. Electrochimica Acta 2004; 49 (22-23) 3807–3820. DOI: 10.1016/j.electacta.2003.12.057

71. McMullan G, Meehan C, Conneely A, Kirby N, Robinson T, Nigam P, Banat IM, Marchant R, Smyth WF. Mini-review: microbial decolorisation and degradation of textile dyes. Applied Microbiology and Biotechnology 2001; 56 (1-2): 81-87. DOI: 10.1007/s002530000587. http://www.springerlink.com/content/04vuqljhajpa32rv/ (accessed 10 April 2012).

72. Pereira WS, Freire RS. Ferro zero: Uma nova abordagem para o tratamento de águas contaminadas com compostos orgânicos poluentes. Química Nova 2005;28(1): 130-136. DOI: 10.1590/S0100-40422005000100022. http://www.scielo.br/scielo.php?script=sci_arttext&pid=S0100-40422005000100022 (accessed 10 March 2012).

73. Houk VS. The genotoxicity of industrial wastes and effluents. Mutation Research 1992; 277 (2) 91-138.

74. Jager I, Hafner C, Schneider K. Mutagenicity of different textile dye products in Salmonella typhimurium and mouse lymphoma cells. Mutation Research 2004;561 (1-2) 35-44.

75. Chequer FMD, Angeli JPF, Ferraz ERA, Tsuboy MS, Marcarini JC, Mantovani MS, Oliveira DP. The azo dyes Disperse Red 1 and Disperse Orange 1 increase the micronuclei frequencies in human lymphocytes and in HepG2 cells. Mutation Research 2009; 676 (1-2) 83-86.

76. Chequer FMD, Lizier TM, Felicio R, Zanoni MVB, Debonsi HM, Lopes NP, Marcos R, Oliveira DP. Analyses of the genotoxic and mutagenic potential of the products formed after the biotransformation of the azo dye Disperse Red 1. Toxicology in Vitro 2011; 25 (8) 2054-2063

77. Ferraz ERA, Umbuzeiro GA, de-Almeida G, Caloto-Oliveira A, Chequer FMD, Zanoni MVB, Dorta DJ, Oliveira DP. Differential Toxicity of Disperse Red 1 and Disperse Red 13 in the Ames Test, HepG2 Cytotoxicity Assay, and Daphnia Acute Toxicity Test. Environmental Toxicology 2011; 26 (5) 489-497

78. Oliveira GAR, Ferraz ERA, Chequer FMD, Grando MD, Angeli JPF, Tsuboy MS, Marcarini JC, Mantovani MS, Osugi ME, Lizier TM, Zanoni, MVB, Oliveira DP. Chlorination treatment of aqueous samples reduces, but does not eliminate, the mutagenic effect of the azo dyes Disperse Red 1, Disperse Red 13 and Disperse Orange 1. Mutation Research 2010 703 (2) 200-208

79. Konstantinou IK, Albanis TA. TiO2-assisted photocatalytic degradation of azo dyes in aqueous solution: kinetic and mechanistic investigations - A review. Applied Catalysis B: Environmental 2004; 49 (1) 1-14. DOI:10.1016/j.apcatb.2003.11.010.

80. Hessel C, Allegre C, Maisseu M, Charbit F, Moulin P. Guidelines and legislation for dye house effluents. Journal of Environmental Management 2007; 83 (2) 171-180. DOI: 10.1016/j.jenvman.2006.02.012.

81. Carneiro PA, Osugi ME, Fugivara CS, Boralle N, Furlan M, Zanoni MVB. Evaluation of different electrochemical methods on the oxidation and degradation of Reactive Blue 4 in aqueous solution. Chemosphere. 2005; 59(3) 431-439.

82. Mohan SV, Bhaskar YV, Karthikenyan J. Biological decolourization of simulated azo dye in aqueous phase by algae Spirogyra species. International Journal of Environment and Pollution. 2004; 21 (3) 211-222.

83. Vaghela SS, Jethva AD, Mehta BB, Dave SP, Adimurthy S, Ramachandraiah G. Laboratory studies of electrochemical treatment

of industry azo dye effluent. Environmental Science and Technology. 2005; 39 (8) 2848-2855.

84. Cardoso JC, Lizier TM, Zanoni MVB. Highly ordered TiO2 nanotube arrays and photoelectrocatalytic oxidation of aromatic amine. Applied Catalysis, B: Environmental. Applied Catalysis, B: Environmental. 2010; 99 (1-2) 96-102.

85. Pereira RS. Identificação e caracterização das fontes de poluição em sistemas hídricos. Revista Eletrônica de Recursos Hídricos. IPH-UFRGS 2004; 1(1): 20-36. http://www.abrh.org.br/informacoes/rerh.pdf (accessed 06 June 2012).

86. Baumgarten, MGZ.; Pozza, SA. Qualidade de Águas. Descrição de Parâmetros Químicos referidos na Legislação Ambiental. Ed: FURG; 2001.

87. Prado AGS, Torres JD, Faria EA, Dias SCL. Comparative adsorption studies of indigo carmine dye on chitin and chitosan. Journal of Colloid and Interface Science. 2004; 1(1) 43-47.

88. Babel S, Kurniawan TA. Low-cost adsorbents for heavy metals uptake from contaminated water: a review. Journal of Hazardous Materials. 2003; 97(1-3) 219-243.

89. Associação Brasileira da Indústria Têxtil e de Confecção. Economia. <http://www.abit.org.br/site/navegacao.asp?id_menu=8&IDIOMA=PT > (accessed 19 August 2011).

90. Walsh FC. Electrochemical technology for environmental treatment and clean energy conversion. Pure and Applied Chemistry. 2001; 73 (12) 1819-1837.

91. Flores ER, Perez F, Torre M. Scale-up of Bacillus thuringiensis fermentation based on oxygen transfer. Journal of Fermentation and Bioengineering. 1997; 83 (6) 561-564.

92. Bornick H. Simulation of biological degradation of aromatic amines in river bed sediments. Water Research. 2001; 35 (3) 619-624.

93. Wang L, Barrington S, Kim JW. Biodegradation of pentylamine and aniline from petrochemical wastewater. Journal of Environmental Management. 2007; 83 (2) 191-197.

94. Orshansky F, Narkis N. Characteristics of organics removal by pact simultaneous adsorption and biodegradation. Water Research. 1997; 31(3) 391-398.

95. Fukushima M, Tatsumi K, Morimoto K. The fate of aniline after a photo-fenton reaction in an aqueous system containing iron(III), humic acid, and hydrogen peroxide. Environmental Science and Technology. 2010; 34 (10) 2006-2013.

96. Pramauro E. et al. Photocatalytic treatment of laboratory wastes containing aromatic amines. Analyst. 1995; 120 (2) 237-242.

97. Augugliaro V. et al. Photodegradation kinetics of aniline, 4-ethylaniline, and 4-chloroaniline in aqueous suspension of polycrystalline titanium dioxide. Research on Chemical Intermediates. 2000; 26 (5) 413-426.

98. Canle LM, Santaballa JA, Vulliet E. On the mechanism of TiO2-photocatalyzed degradation of aniline derivatives. Journal of Photochemistry and Photobiology, A: Chemistry. 2005; 175 (2-3) 192-200.

99. Chu W, Choy WK, So TY. The effect of solution pH and peroxide in the TiO2-induced photocatalysis of chlorinated aniline. Journal of Hazardous Materials. 2007; 141 (1) 86-91.

100. Low, G. K. C.; McEvoy, S. R.; Matthews, R. W. Formation of nitrate and ammonium ions in titanium dioxide mediated photocatalytic degradation of organic compounds containing nitrogen atoms. Environmental Science and Technology, v. 25, n. 3, p. 460-467, 1991.

101. Pinheiro HM, Touraud E, Thomas O. Aromatic amines from azo dye reduction: status review with emphasis on direct UV spectrophotometric detection in textile industry wastewaters. Dyes and Pigments. 2004; 61(2) 121-139.

102. Fukushima M, Tatsumi K, Morimoto K. The Fate of Aniline after a Photo-Fenton Reaction in an Aqueous System Containing Iron(III), Humic Acid, and Hydrogen Peroxide. Environmental Science and Technology. 2006; 34 (10) 2006-2013.

103. Chu W, Choy WK, So TY. The effect of solution pH and peroxide in the TiO2-induced photocatalysis of chlorinated aniline. Journal of Hazardous Materials. 2007; 141 (1) 86-91.

104. Finklea HO. Semiconductor electrodes. New York: Elsevier; 1998.

105. Lin SH, Chen ML. Purification of textile wastewater effluents by a combined Fenton process and the ion exchange. Desalination 1997; 109 (1) 121-130.

106. Mozia S. Photocatalytic membrane reactor (PMRs) in water and wastewater treatment. A review. Separation and Purification Technology 2010; 73 (2) 71-91.

Chapter 2

Textile Dyeing: Environmental Friendly Osage Orange Extract on Protein Fabrics

Heba Mansour[1, 2]

[1] *King Abdul- Aziz University, Faculty of Arts and Design, Fashion Design, Saudi Arabia*

[2] *Helwan University, Faculty of Applied Arts, Textile printing dyeing and Finishing Department,*

1. INTRODUCTION

Dyeing was known as early as in the Indus Valley period (2600-1900 BC); this knowledge has been substantiated by findings of colored garments of cloth and traces of madder dye in the ruins of the Indus Valley Civilization at Mohenjodaro and Harappa. Natural dyes, dyestuff and dyeing are as old as textiles themselves. Man has always been interested in colors; the art of dyeing has a long past and many of the dyes go back into prehistory. It was practiced during the Bronze Age in Europe. The earliest written record of the use of natural dyes was found in China dated 2600 BC. [1-4] Primitive dyeing techniques included sticking plants to fabric or rubbing crushed pigments into cloth. The methods became more sophisticated with time and techniques using natural dyes from crushed fruits, berries and other plants, which were boiled into the fabric and which gave light and water fastness (resistance), were developed. After the accidental synthesis of mauveine by Perkin in Germany in 1856 and its subsequent commercialization, coal-tar dyes began to compete with natural dyes. With advances in chemical techniques, the manufacture of synthetic dyes became possible, leading to greater production efficiency in terms of quality, quantity and the potential to produce low-cost raw materials. As a result, natural dyes were progressively replaced by synthetic dyes, whereas over 80% of which is constituted of the aromatic azo type [5]· However, researches have shown that synthetic dyes are suspected to release harmful chemicals that are allergic, carcinogenic and detrimental to human health. In addition, textile industries produce huge amounts of polluted effluents that are normally discharged to surface water bodies and ground water aquifers. These wastes cause many damages to the ecological system of the receiving surface water, creating a lot of disturbance to the ground water resources [6-9]. Therefore, in 1996, ironically Germany became the first country to ban certain azo dyes [10].

The world-wide demand for fibers and safety dyes is increasing probably according to the greater awareness of the general consumers in the USA, Europe and Japan towards the highly pollutant procedures affecting fiber and textile coloration when using synthetic dyes which act as a sources of skin cancer, disorders and allergic contact dermatitis [11]. Therefore, interest in returning back to natural dyes as synthetic dyes substitute has increased considerably on account of their high compatibility with environment, relatively low toxicity and allergic effects, as well as availability of various natural coloring sources such as from plants, insects, minerals and fungi [12]

Natural dyes can be obtained from plants, animals and minerals, producing different colors like red, yellow, blue, black, brown and a combination of these. For technical application of natural dyes, a number of requirements have to be fulfilled;

Most problems are derived from technical demands, for example:

- adaptation of traditional dyeing processes on modern equipment [13]
- supply of dye-houses with the required amount of plant material [13]
- standardization of extraction and dyeing of the plant material [14]
- selection of plant material and processes that yield products with acceptable fastness properties [14]

From an industrial point of view it would be easier to resort to extracts despite there is at present no definite answer to this prospective solution. The simplest extract would be a watery one although not all the dye pigments are water-soluble. Use of organic solvents might give rise to extracts which are not completely water-soluble [15], provided that the solvent chosen guarantees a series of properties as follows:

- Its extraction capacity is extremely high for practically all the natural pigments present in the raw materials of interest.

- Its boiling temperature and latent heat of vaporization is quite low to allow its separation at low temperatures with minimum energy consumption;
- Its reactivity with colors and pigments is insignificant to avoid any loss in the color quality [15].

Natural dyes are substantive and require a mordant to fix to the fabric, and prevent the color from either fading with exposure to light or washing out. These compounds aid a chemical reaction between the dye and the fiber, so that the dye is absorbed. Traditionally, mordants were found in nature. Wood ash or stale urine may have been used as an alkali mordant, and acids could be found in acidic fruits or rhubarb leaves.

There are three types of mordant: i) Metallic mordants; Metal salts of aluminium, chromium, iron, copper and tin, ii) Tannins; Myrobalan and sumach, iii) Oil mordants; mainly used in dyeing Turkey red color from madder by forming a complex with alum.

In order to obtain high color yield, different shades and good fastness properties, metallic salt mordants are normally employed [16]. The application of mordants for dye fixation was carried out by three methods; i- Pre-mordanting; dipping the fabric in the mordant solution before dyeing, ii- Simultaneous mordanting; addition of mordant in the dye bath during the dyeing, and iii- post mordanting; dipping the fabric in the mordant solution after dyeing [17].

Osage orange (Maclura pomifera) is a tree in the Moraceae family [18]. The common name is derived from its fruit, which resembles the shape of an orange, and from the fact that its hardwood was used by the Osage Indian tribe to make bows. It is native to Southern Oklahoma and Northern Texas, and is planted throughout the United States. Several compounds have been isolated and identified in various parts of this tree namely, isoflavonoids from the fruit, flavonols and xanthones from the heartwood and stem bark, and flavanones and xanthones from the root bark [19] It contains lectins, triterpenes, xanthones and flavone-type compounds such as Scandenone and auriculasin as shown in Figure 1 [20] Two predominant isoflavones, pomiferin and osajin, are derived from the simple isoflavone genistein by prenyl substitutions [21] as shown in Figure 2.

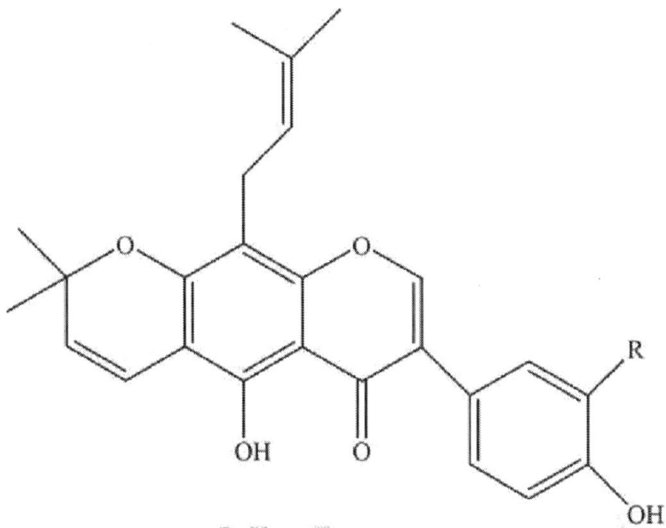

R=H (I)
R=OH (II)

FIGURE 1. Chemical structures of scandenone (I) and auriculasin (II)

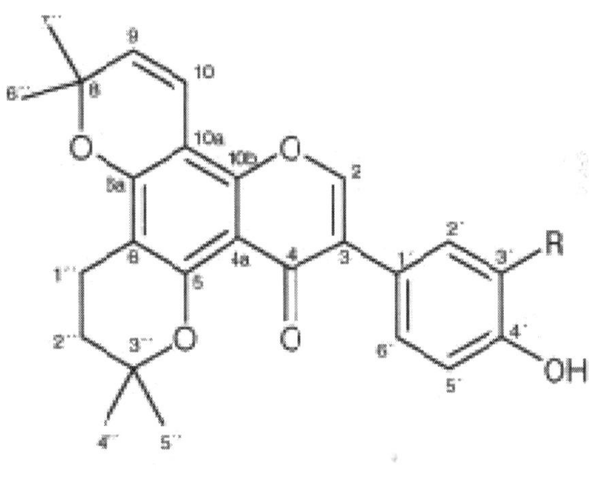

i R= H (*Iso*-Osajin)
ii R= OH (*Iso*-Pomiferin)

FIGURE 2. Chemical structures of Osajin (i) and pomiferin (ii)

The compounds, osajin, iso-osajin, pomiferin, and iso-pomifcrin have been characterized and their chemical structures determined. They are isoflavones with the following structures as shown in Figure 3. [19, 21].

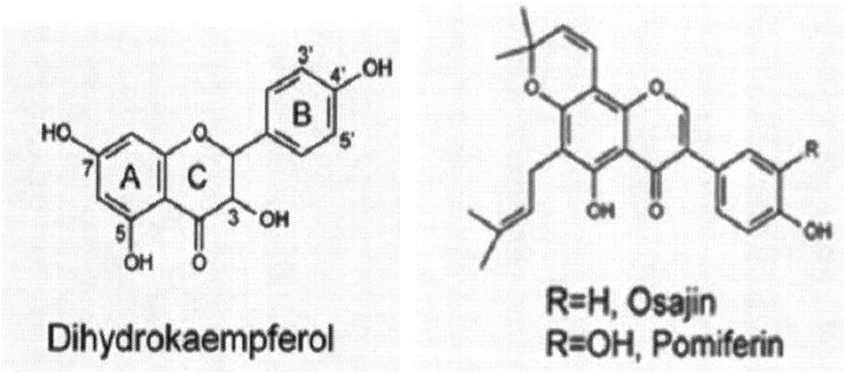

FIGURE 3. Chemical Structure of iso-pomifcrin

As well as Osage orange was applied as an eco-friendly dye acting as one of the environmental problems solutions. New concepts in the cleaner production are being evaluated to solve the high water and energy consumption in textile industries.

The use of ultrasonic as a renewable source of energy in textile dyeing has been increased due to many advantages associated with it [22-25]. Ultrasonic energy represents a promising technique for assisting silk treatment, dyeing, and mordanting processes in comparison with the conventional heating technique. Sonic energy succeeded in accelerating the rate of, dyeing, and mordanting at lower temperatures rather than the conventional heating technique

Therefore, the present investigation was aimed at identifying the most appropriate leaching solvent for Osage orange pigments to produce an optimum concentrated extract used for dyeing protein fibers; silk and wool fabrics. This has been carried out ultrasonically in comparison with the classical thermal method, using water, in addition to the co-solvents of water-acetone, and water-ethanol mixtures at different concentrations, temperature and time intervals. The optimum condition of the efficiency of ultrasonic assisted dyeing and mordanting methods of Osage orange extraction on the quality of the dyed protein based materials were determined

2. EXPERIMENTAL

2.1. Materials

Degummed and bleached plain Habotain silk fabric, and 100% mill scoured wool fabric purchased from Sherazad Com. New Zealand, were further washed with a solution containing 0.5 g/L of sodium carbonate and 2 g/L of non-ionic detergent (Nonidet ® P 40 Substitute purchased from *Sigma- Aldrich NZ. Ltd.*), keeping the material to liquor ratio at 1:50, for 30 min. at 40-45°C. The scoured materials were thoroughly washed and dried at ambient temperature. [23]

An analytical grade alum and the commercially cream of tartar were used as mordantas.

2.2. Pigment Leaching And Estimation Of Extraction Yields

To select the best solvent for Osage orange, distilled water and other co-solvents, such as water-acetone, and water- ethanol mixtures all of analytical grade, were tested at concentrations of 10% v/v. 2 g of Osage orange powder (Hands Ashford NZ, LTD, ChristChurch, NZ) was suspended in 20 $cm^3$of solvent, and in thermostatic as well ultrasonic baths at 60 °C, for 120 min. Once water-acetone co-solvent, and ultrasonic assisted extraction were chosen as the preferable technique of extraction, 10 % w/v Osage orange powder, dissolved in (2.5- 25) % v/v acetone, at (25- 60) °C, for 30- 120 min, were carried out to determine the standardization method of extraction. [24]

2.3. Dyeing and mordanting

1 % Osage orange extract was filtered and used as a dyeing bath. Silk and wool samples were added to the extract, and the dyeing parameters were studied ultrasonically keeping the material to liquor ratio at L: R of 1:30, for time intervals varied between 30- 120 min, at temperatures from 30-60°C. In terms of the pH used for dyeing; the pH values ranging from (3-11) were carried out to control the dye uptake.

On studying the mordanting methods, the optimum concentration of 8% w/v Osage orange extract was carried out. Stock solutions of 50 gm/ l alum and 25 g/l mixture of each of alum and cream of tartar were prepared. Two different methods of mordanting were used: (1) pre-mordanting method: the samples were first mordanted and then dyed without intermediate washing; and (2) post-mordanting method: the samples were first dyed and then mordanted. Ultrasonic assisted mordanting was carried out in comparison with the conventional heating method at 50-60°C, for 90 min, at pH 5. samples were rinsed, washed with 0.5 g/L sodium carbonate and 2 g/L of non-ionic detergent at 40-45°C for 30 min, keeping the material to liquor ratio at 1:50. Finally washed with water, and dried at ambient temperature. [23, 26, 27]

2.4. Measurements

Dyestuff content of the dyed fabrics was determined according Kubelka–Munk equation [28] using*Cary 100 UV-Vis Spectrophotometer*

$$f(R)=\frac{(1-R)^2}{2R}=\frac{k}{s}$$

R is the absolute reflectance of the sampled layer, K is the molar absorption coefficient and s is the scattering coefficient.

After which the samples were tested for color fastness to light and washing according to AATCC107-1997 [29] The CIE-Lab values of the dyeings were measured and the cylindrical co-ordinates of color were determined after exposure to arc lamp irradiation for 1, 2, 4, 6, 24, 48, and 72 hrs. The colors are given in an internationally commission (CIE L*a*b*) coordinates, L* corresponding to brightness, a* to red–green coordinate (positive sign = red, negative sign = green) and b* to yellow–blue coordinate (positive sign = yellow, negative sign = blue). [30-32].

$$\Delta E^*_{ab} = \{(\Delta L^*)^2 + (\Delta a^*)^2 + (\Delta b^*)^2\}^{1/2}$$

3. RESULTS AND DISCUSSION

3.1. Extraction Parameters

Extraction was carried out by water, water-acetone, and water-ethanol mixtures at 25-60 ◦C for 30-120 min. The ultrasonic efficiency had been determined simultaneously with extraction parameters, and compared with the conventional heating method. The yield coefficients of co-solvents were definitively greater than water, with much higher values in case of ultrasonic. Water-acetone mixture was found to be the most selective co- solvent followed by water-ethanol. As shown in Figures [4, 5] water-acetone mixture released over 32% of the total dye absorbency, exhibiting 21% of the total color strength when dyeing the woolen sample. Water-ethanol extracted 27% dye and exhibited 19% color strength, while water extracted less than 21% dye, and exhibited 16% color strength. This was relative to [10, 6, and 4] % of absorbance and [18, 15, and 11] % color strength respectively with the co-ordinate solvents when using the conventional heating method.

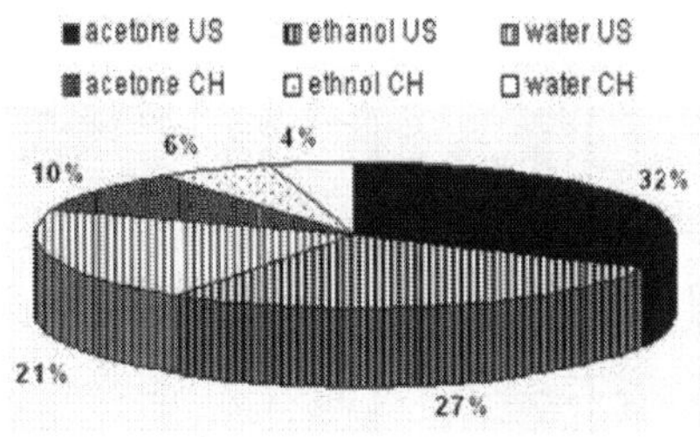

FIGURE 4. Effect of solvents on the absorbency of Osage orange powder using the conventional [CH] and ultrasonic [US] assisted extraction

Figures 6, 7 and 8 showed the absorption, and color strength values of Osage orange powder extracted by acetone at different concentrations of (2.5 – 25) % v/v at temperatures from 25-60° C, for time intervals varied between 30-120 min.

The maximum values were achieved with 20 % v/v acetone at 60° C, for 60 min. The extraction parameters affected the color strength and are influenced by the properties of solvents such as, the dipole moment, dielectric constant, and refractive index values.

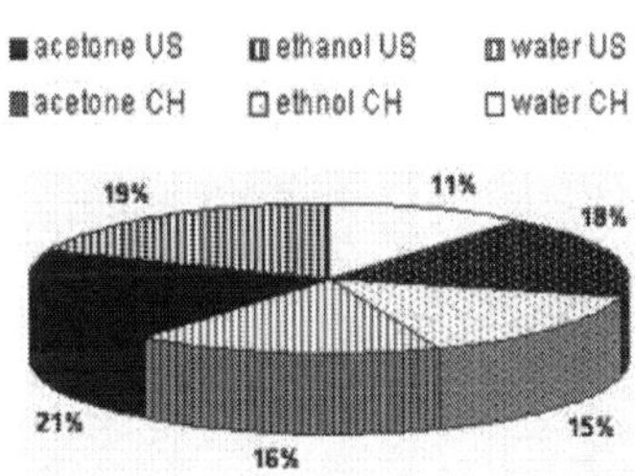

FIGURE 5. Effect of solvents on the color strength of Osage orange dyed wool using the conventional [CH] and ultrasonic [US] assisted extraction

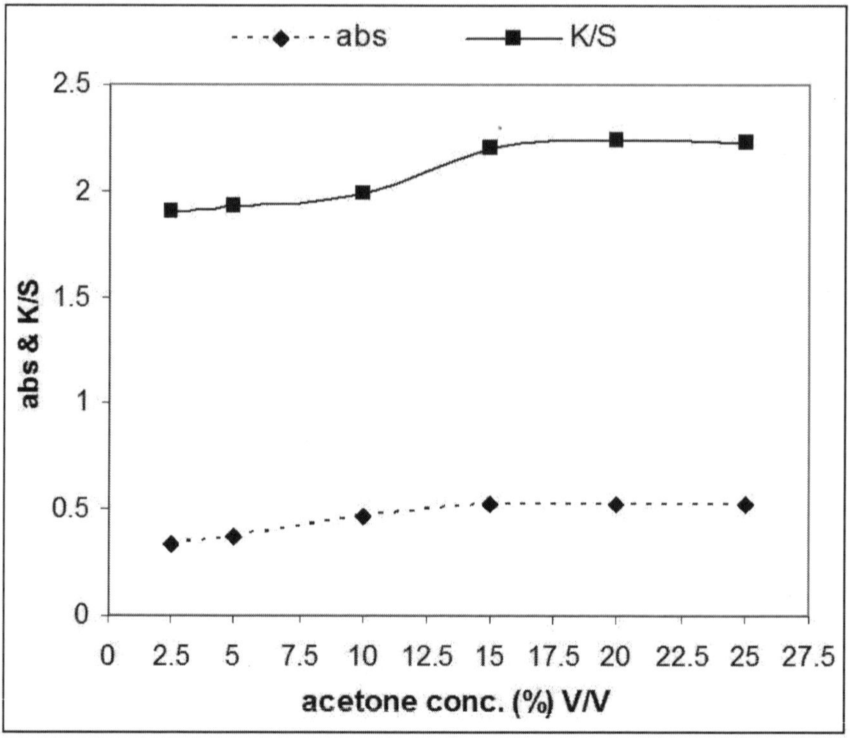

FIGURE 6. Effect of acetone concentration on the absorbency and color strength of Osage orange dyed wool using ultrasonic assisted extraction

The solvent polarity can change the position of the absorption or emission band of molecules by solvating a solute molecule or any other molecular species introduced into the solvent matrix [33]. By the way, dye molecules are complex organic molecules which might carry charge centers and are thus prone to absorption changes in various media [33, 34]

Acetone acts as the non hydrogen-bond donating solvents (also called as non-HBD type of solvents), while water and ethanol are the hydrogen-bond donating solvents (also called as HBD type solvents) [33]. The absorbency values of Osage orange in these solvents are given in Figure 4, 5. It can noted from this figure that the absorption maximum of the extract is affected by the solvent type, thus the change in values can be noted as a probe for various types of interactions between the solute and the solvent.

Water and ethanol are considered as polar protic solvents, their polarity stems from the bond dipole of the O-H bond, whereas the large difference in the electro- negativities of the oxygen and hydrogen atom, combined with the small size of the hydrogen atom, warrant separating the Osage orange molecules that contain the OH groups from those polar compounds that do not. On the other hand acetone considered as dipolar aprotic solvent, containing a large multiple bond between carbon and either oxygen or nitrogen e.g. C-O double bond. [33-35].

Although water has the highest dielectric constant among ethanol and acetone solvents, its extraction demonstrated the lowest value of absorbency. This might due to the formation of strong hydrogen bond between the dyes extract and water molecules [33-35].

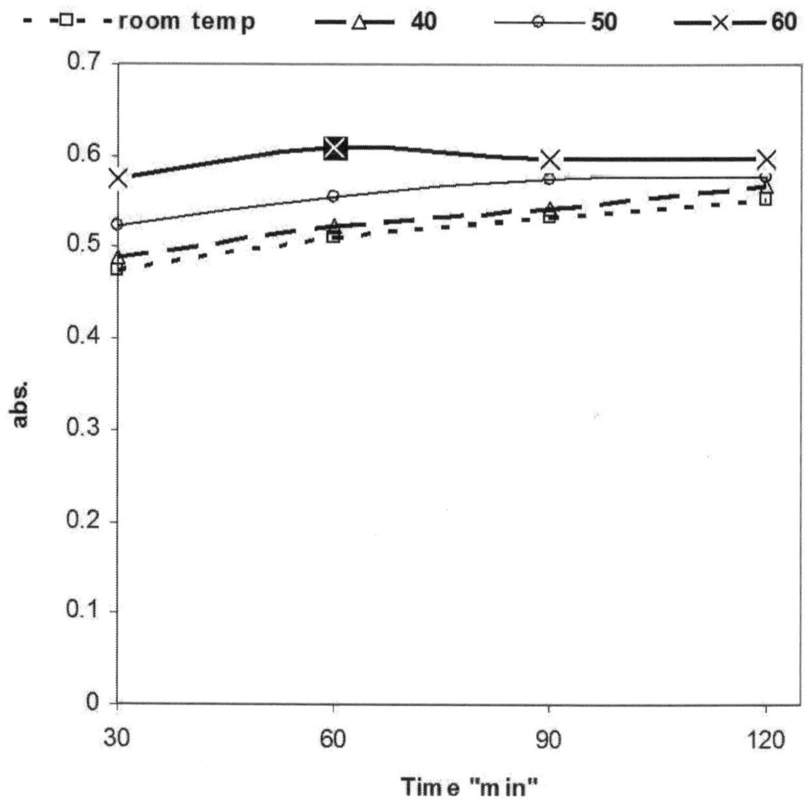

FIGURE 7. Effect of ultrasonic assisted extraction time on the absorbency of Osage orange at different temperatures

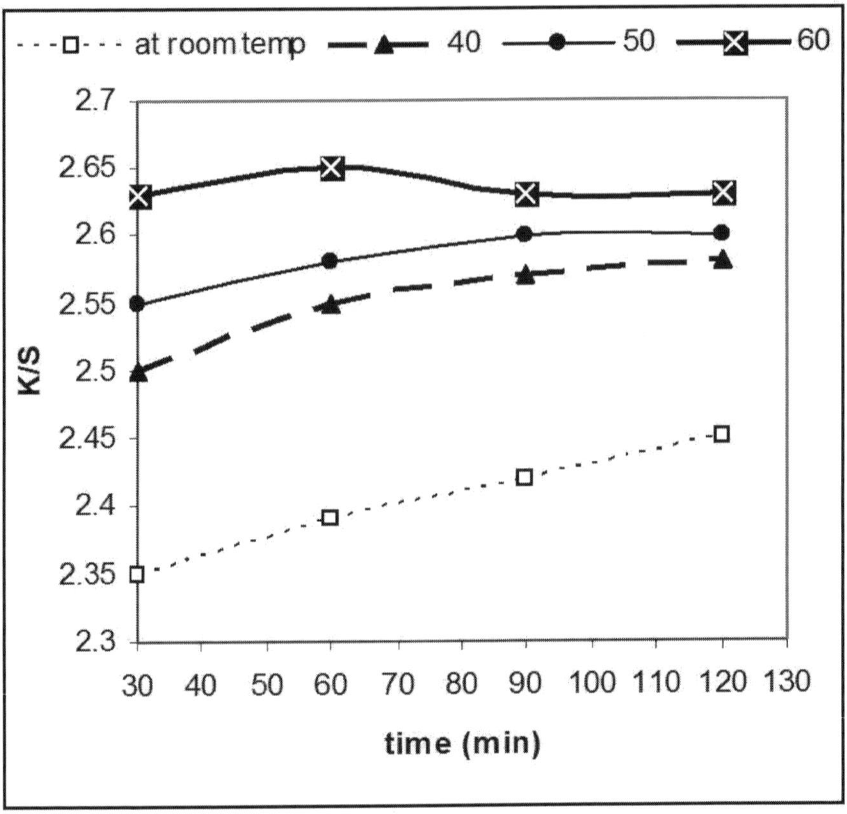

FIGURE 8. Effect of ultrasonic assisted extraction time on the color strength of Osage orange dyed wool at different temperatures

The dye absorbance is also influenced by the presence of co-solvents. Water-acetone mixture exhibited the highest value of absorbance, followed by the second water-ethanol mixture. In case water-acetone, the salvation of extract is non-HBD type of solvent mainly occurs through charge-dipole type of interaction, whereas in HBD type of solvent, the interaction also occurs by hydrogen bonding besides the usual ion-dipole interaction. In this situation, the methyl groups of acetone are responsible for the solvation of the dye extract. Thus, decreasing the amount of non-HBD acetone solvent "concentration" increasing the amount of HBD solvent (water) shall break these interactions with the dye molecule, thereby decreasing the value of absorbance. Water-ethanol mixtures belong to HBD type of solvents, whereas the dye cation is preferentially solvated by the alcoholic component in all mole fractions in aqueous mixtures with ethanol. It is well known that water makes strong hydrogen-bonded nets in the water-rich region, which are not easily disrupted by the co-solvent [33, 34]. This can explain the strong preferential salvation by the alcoholic component in this region since water preferentially interacts with itself rather than with the dye. In the alcohol-rich region, the alcohol molecules are freer to interact with the water and with the dye, since their nets formed by hydrogen bonds are weaker than in water. In this situation, the alcohol molecules can, to a greater or lesser extent, interact with water through hydrogen bonding [33-35]

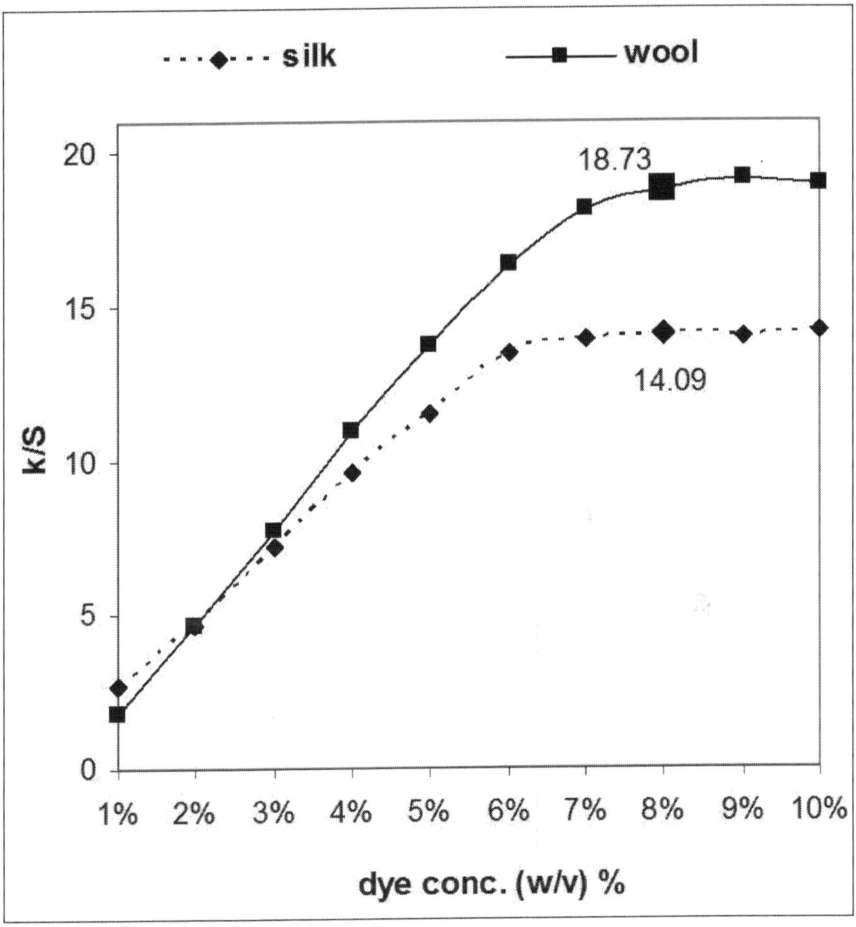

FIGURE 9. Effect of Osage orange concentration (w/v)%, extracted ultrasonically in 20 % water/ acetone co-solvent at 60 °C for 90 min, on the color strength of silk and wool samples.

Wool fiber is considered as relatively easy fiber to dye, the ease with which the polymer system of wool will take in dye molecules is due to polarity of its polymer and its amorphous nature. The polarity will readily attract any polar Osage orange molecules and draw them into the polymer system. The studies of wool dyeing process have been in two distinct theories (The Gilbert- Rideal's and Donnan theories). The Gilbert and Rideal theory based on Langmuir's theories of surface adsorption [36], in which the activity coefficient of Osage orange extract ions adsorbed into the wool phase are reduced due to specific binding with sites on wool, which is the formation of ion pairs. This theory proposed that dyeing process is an anion exchange process, in which the Osage orange extract molecules displace smaller anions, depending on four steps: a) diffusion to fiber surface, b) transfer across that surface, c) diffusion within to appropriate "sites" and d) binding at those sites. On the other hand, according to the Donnan equilibrium theory, the Osage orange extract was considered to partition between the external solution and internal solution phase in the wool. The later phase is believed to contain a high concentration of fixed ionic groups, and hence solute molecules have reduced activity co-efficient in that phase due to coulombic interaction between the anionic groups (OH) in fact: O^- of Osage orange extract and the protonated amino groups of wool. [36]

Higher color depth was expected from an increase in the extract concentration and the use of high concentrations of mordant [37]

To study the possibility of forming concentrated extracts, different amounts of Osage orange bark powder (1-10) g were extracted per the optimized 20 % water/ acetone co-solvent. The dyeing process was carried out on silk and wool samples at a liquor ratio of L: R 1:20, for 60 min. at 60 °C. It was noted that, the use of more concentrated extracts resulted in somewhat an increase in color depth; whereas the maximum color depth was achieved with 8% w/v powder on both fabrics as shown inFigure 9. The relative high K/S values for dyeings can be explained with the high amount of bark extracted for this series of dyeing experiments.

3.2. Dyeing Parameters

Dyeing temperature and time are important parameters influencing the quality of the dyed silk and wool samples. It is well known that dyeing at high temperature for a long time tends to decrease the fabric strength. [36-38] Therefore, it was proffered to dye the samples ultrasonically at temperatures ranging from 30 to 60 °C, relative to the dyeing time that was studied from 30 to 120 mins.

As shown in Figures 10 and 11, it is clear that the standard parameters of dyeing temperature and time were achieved after 90 mins, at 40 – 50 ° C, and 50-60 ° C in case of silk and wool respectively, where the color strength increases with the increase in dyeing temperature.

Generally, the increase in dye-uptake can be explained by the fiber swelling which enhanced the dye diffusion. [37] The effect of dyeing time was conducted to reveal the effect of power ultrasonic on the de-aggregation of dye molecules in the dye bath. It was denoted that the color strength obtained increased as the time increased. The decline in the dye-ability may be attributed to the hydrolytic decomposition of the extract molecules under the influence of sonic energy during prolonged dyeing. [38]

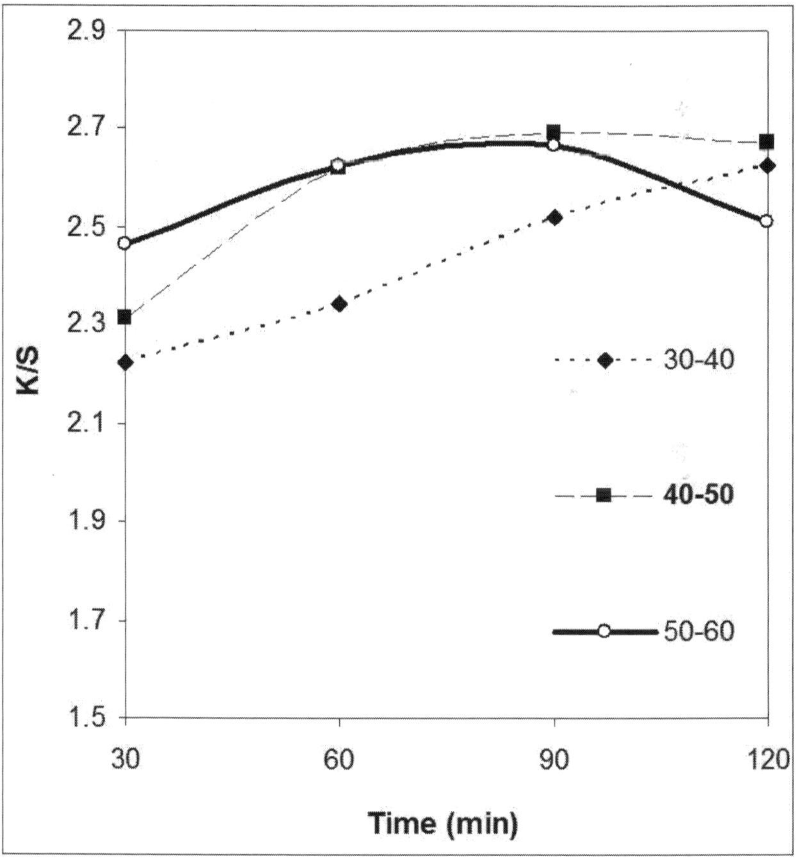

FIGURE 10. Effect of dye bath temperature at different time intervals on the color strength of silk samples dyed ultrasonically with 1% (w/v) Osage orange extract.

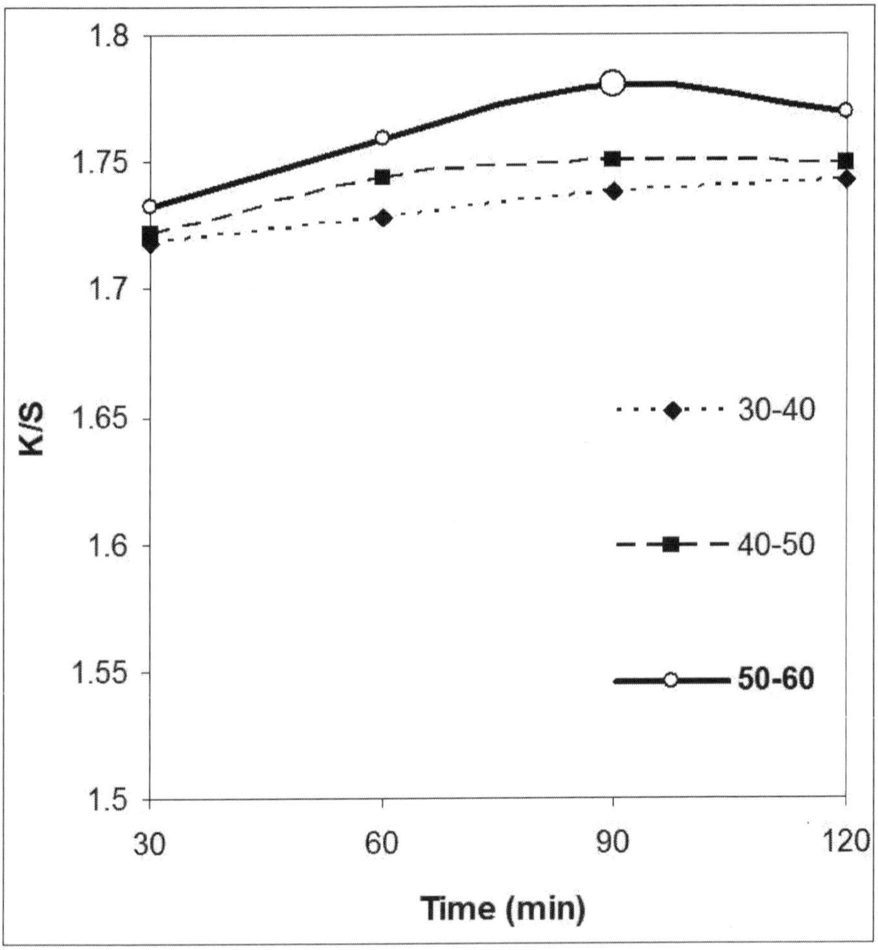

FIGURE 11. Effect of dye bath temperature at different time intervals on the color strength of wool samples dyed ultrasonically with 1% (w/v) Osage orange extract

As shown in Figures 12 and 13 the pH values of the dye bath, have a considerable effect on the dyeability of silk and wool fabrics with Osage orange extract under ultrasonic. As the pH increases the dyeability, decreases. The effect of dye bath pH can be attributed to the correlation between dye structure and the protein based materials.

Since the used dye is a water-soluble dye containing hydroxyl groups, it would interact ionic-ally with the protonated terminal amino groups of silk and wool fibers at acidic pH via ion exchange reaction.

The anion of the dye has a complex character, and when it is bound on fiber, further kinds of interactions take place together with ionic forces. This ionic attraction would increase the dye-ability of the fiber as clearly observed in Figures 12 and 13. At pH greater than 5, the ionic interaction between the hydroxyl anion of the dye and the protein fibers decreases due to the decreasing number of protonated terminal amino groups of silk and wool and thus lowering their dye-ability. It is to be mentioned that the lower dye-ability may be attributed to the enhanced desorption of the dye as its ionic bond is getting decreased [36].

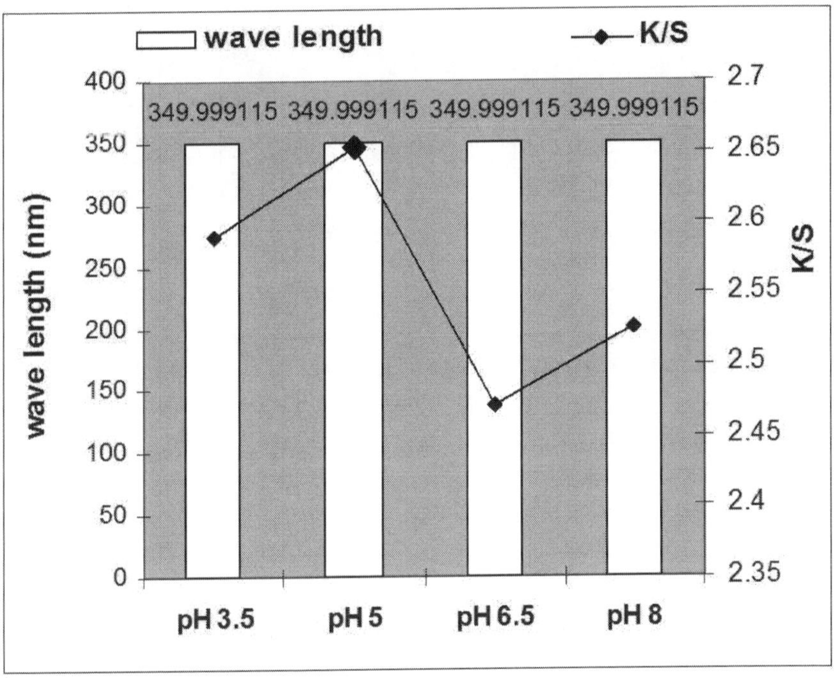

FIGURE 12. Effect of the dye bath pH on the color strength and wave length of silk samples dyed ultrasonically with 1% Osage orange at 40-50°C for 90 min.

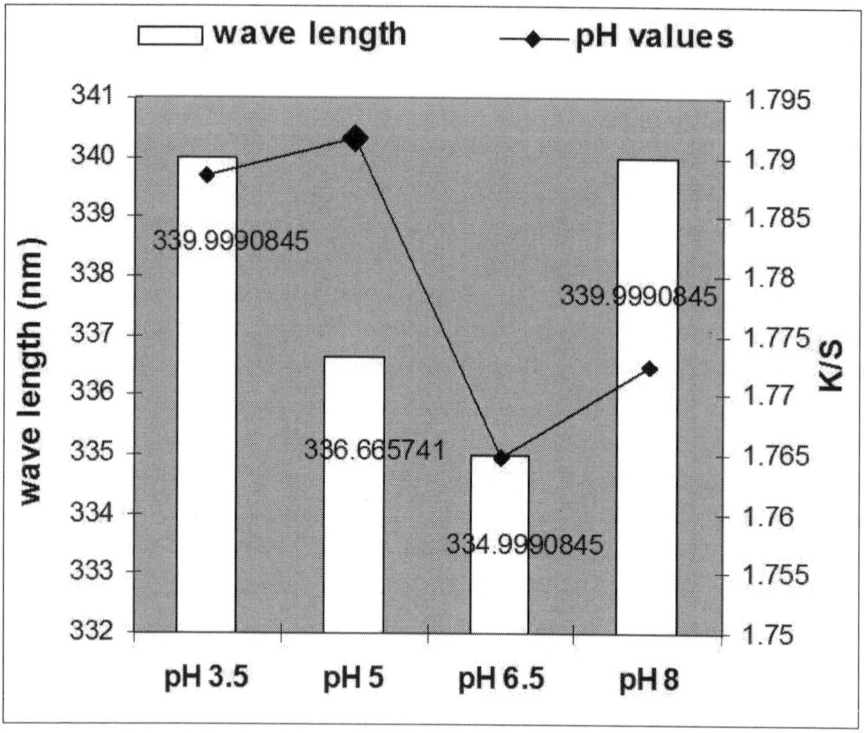

FIGURE 13. Effect of the dye bath pH on the color strength and wave length of woolen samples dyed ultrasonically with 1% Osage orange at 50-60 °C for 90 min.

3.3. Mordanting Methods & Colour Properties

As shown in Figures 14 and 15, ultrasonic [US] assisted mordanting method possesses a remarkable improvement in the color strength, in comparison with the classical thermal method [CH]. The obtained dyeings are governed by the descending dyeing sequence, and can be ranked as follows: pre-mordanting with a mixture of alum and cream of tartar followed by dyeing > post mordanting with alum > post mordanting with a mixture of alum and cream of tartar > premordanting with alum > unmordanted samples.

Silk and wool fabrics are highly receptive to mordants due to their amphoteric nature; they can absorb acids and bases with equal effectiveness. Mordants during natural dyeing, exhibits fast color due to their complex formation with the dye and fiber [27, 38]

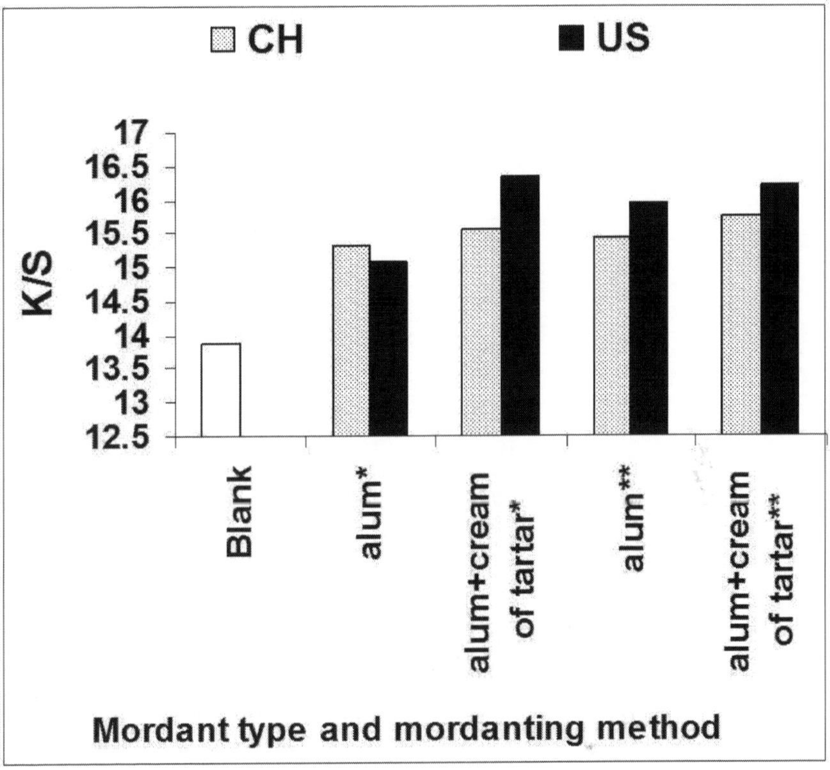

FIGURE 14. Effect of mordant and mordanting method on the color strength of silk dyeings

It is clear that: (i) pre-mordanting with the nominated mordant brings about a significant enhancement in the K/S values of the obtained dyeings. (ii) the extent of improvement is governed by the physical and chemical states of the dye and degree of fixation, (iii) premordanting followed by dyeing gives dyeings with better fastness properties than those dyed without mordant and mordanting after dyeing (iv) the improvement in the dyeings color strength and fastness, reflects higher extent of dye adsorption, interaction and bridging with the pre-mordanted substrate via different conjugated bonds [27]

The low color strength in post-mordanting condition is due to the accumulation of the metal dye complex in form of clusters. [39]. The high aluminum content might provide useful eco-friendly chelating with Osage orange molecules presented in the extract that might resist their hydrolysis by water. [39]

The commercial cream of tartar (Potassiun Bitartrate) contains a small percentage of calcium tartrate is frequently employed as a mordant for wool. [40]. In this study it was recommended to apply cream of tartar with alum as preferable mordant to get good strong colors as discussed previously. It helps to soften fibers when alum is used, and can also help brighten the yellow color with good levelness. [40]

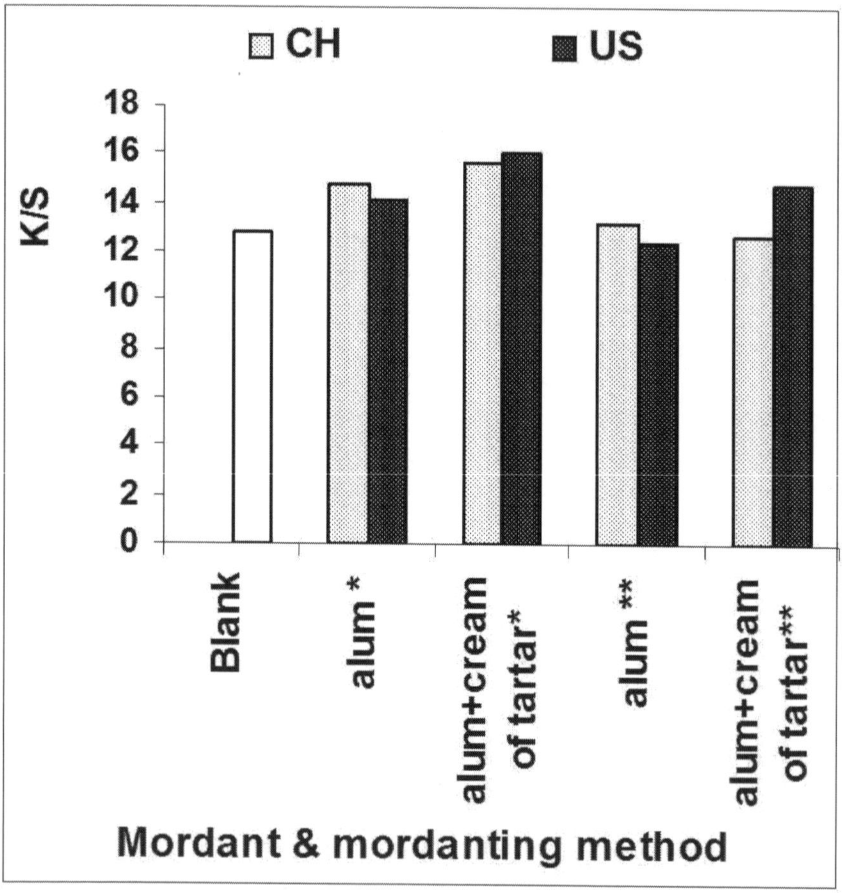

FIGURE 15. Effect of mordant and mordanting method on the color strength of wool dyeings

Sonic energy succeeded in accelerating the rate of mordanting at lower temperatures rather than the conventional heating technique. The exhibited improvement was generally attributed to the acoustic cavitation, which is the formation of gas-filled micro-bubbles or cavities in a liquid media, producing implosive collapse, which often forming fast-moving liquid jets, where large increases in temperature and pressure are generated. The micro-jets increase the diffusion of solute inside the intermediate spaces of silk and wool fabrics facilitate the de-aggregation of Osage orange molecules in the dye bath and thus increase the dye diffusion rate, and penetration through the fibers. [22-24, 41]

A variety of color hues were obtained with respect to the mordant. It was observed from the color fastness data that the extracted dye from *Osage orange* furnished different color hues with very good affinity for silk and wool fabrics in presence of alum and cream of tartar mordant as illustrated inTables 1 and 2. The color intensity reached its highest value when the fabrics treated with a mixture of alum and cream of tartar. The brightness of the shades on the dyed samples might be due to the better absorption of Osage orange extract and the easy metal complex formation of mordant with the fibers. Data represented good to very good fastness to washing, because the mordant lead to the formation of a complex dye which aggregates the dye molecules into a large particles insoluble in water. Control samples exhibited poor fastness to washing due to the weak dye- fiber bond, and the ionization of the (OH) groups of the dye during washing under the alkaline condition. [42-45]

TABLE 1. The CIELab values and fastness properties of Osage orange dyed silk

Sampls	L^*	a^*	b^*	Wash fastness	Light fastness
Blank US	54.12	4.71	49.2	3	5
alum* CH	50.95	3.57	43.37	3-4	5-6
alum* US	55.05	3.56	51.67	4	6
Alum + cream of tartar * CH	50.57	3.81	37.31	4	6-7
Alum + cream of tartar * US	50.29	6.43	44.8	4	6-7
alum** CH	54.45	4.83	48.08	3-4	6
alum** US	54.99	5.40	52.12	3-4	6
Alum + cream of tartar ** CH	53.01	3.96	38.21	3-4	6
Alum + cream of tartar ** US	52.75	5.52	47.49	3-4	6

TABLE 2. The CIELab values and fastness properties of Osage orange dyed wool

Sample	L^*	a^*	b^*	Wash fastness	Light fastness
Blank US	68.97	- 0.02	55.19	3	5
alum* CH	65.07	- 0.59	60.08	3-4	5-6
alum* US	67.46	- 2.93	62.40	4	6
Alum + cream of tartar * CH	65.43	- 0.14	56.89	4	6-7
Alum + cream of tartar * US	64.11	- 1.14	52.60	4	6-7
alum** CH	64.61	0.68	62.40	3-4	6
alum** US	64.71	1.56	58.98	3-4	6
Alum + cream of tartar ** CH	64.54	0.40	60.72	3-4	6
Alum + cream of tartar ** US	65.03	1.23	55.29	3-4	6

In the light fastness test, mordanted colored samples exhibited better light fastness relative to the control ones. This may be due to: i) the aluminum metal in alum mordant protects, both by the steric and electronic effects of the weak point in the dye structure from attack by means of the reactive species during photochemical reaction, in addition ii) the aluminum metal in alum mordant promotes aggregation of the dye. By the way, the poor light fastness is due to the inherent susceptibility of the dye chromophore to the photochemical degradation. [46-51]

The colorimetric data indicated the depth and natural tone of the control and mordanted dyed samples. The L^* values were found to be lower using alum and cream of tartar as mordant corresponding to deeper shades. The L^* values were found to be higher in case of unmordanted dyed samples corresponding to lighter shades. Similarly, by using alum as mordant the L^* values were also higher corresponding to lighter shades. The higher values of a^* and b^* indicated the brightness, representing the redness and yellowness hues respectively. As a result, alum and cream of tartar might be effectively used as mordant for Osage orange extract.

The light fading of the dyed samples was recorded in terms of the color difference *(ΔE)* as shown inFigures 16-20. It was denoted that sonic energy assisted alum and cream of tartar mordanting method, exhibited a lower degree of fading in comparison with the conventional heating and the application alum mordant in absence of cream of tartar, whereas the pre-

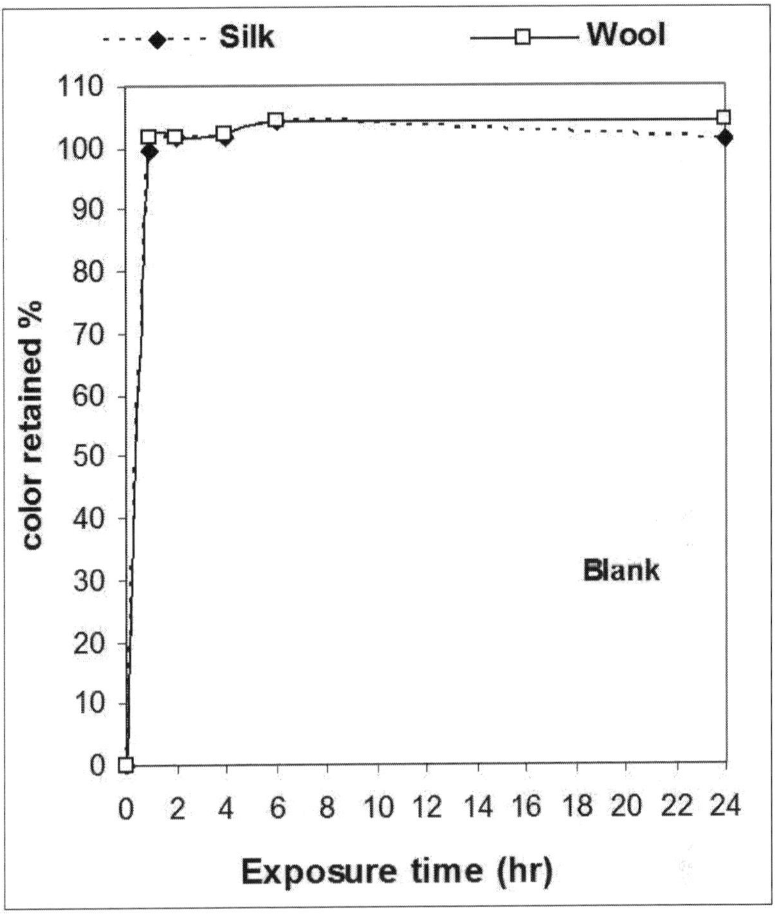

FIGURE 16. Effect of Xenon arc lamp exposure time on the color retained of silk and wool blank dyed samples

mordanting method appears to be preferred having a great efficiency in lowering the degree of fading in comparison with the post mordanting

The dye physical state is generally important than the chemical structure in determining the color fastness on fibers. (31). It was recognized that the light fastness of many dyed systems has been found to increase with the increase dye concentration applied to the substrate [50].

The fiber swelling was increased with ultrasonic technique due to the sonic energy which i) improves the diffusion and penetration of the dye and mordant molecules inside the pores of the fabric, and ii) the fast breaking-down of the

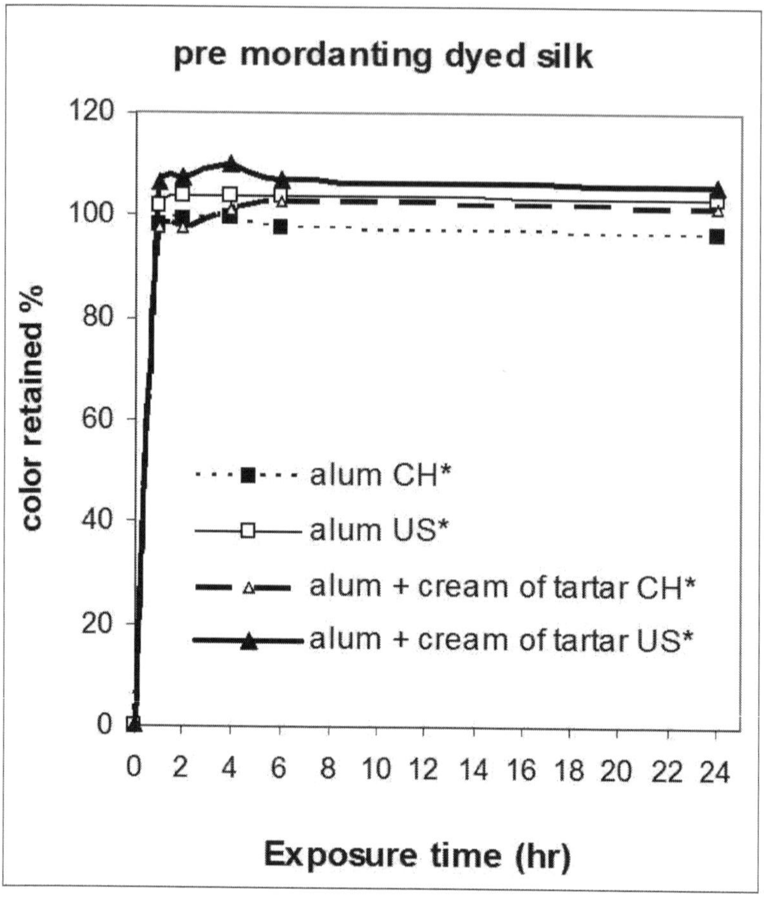

FIGURE 17. Effect of Xenon arc lamp exposure time on the color retained of the premordanted dyed silk samples

dye molecules, which became much more smaller in size and thus fully dispersed with much higher amount in the dyed samples relative to the samples subjected to the conventional heating method, resulted in, the lower degree of fading in case of ultrasonic assisted dyeing and mordantinting processes. [22-25, 41]

Mixture of alum and cream of tartar mordant renders the dye more bonded and more aggregated onto fibers, therefore the surface area of the dye accessible to light is reduced, and thereby the dye fades at lower degree with nearly constant rate of fading.

Aluminum ions apparently produce metal chelates with improved the overall fastness properties. This either could be evidence of the aggregation of dye molecules within the fiber or perhaps of the formation of dye-metal chelates that forms grater stability of the dye molecules when co-ordinated with the complex aluminum metal atom that might form quite large aggregates giving the highest light fast with the lowest fading degree. [23, 51], resulted in, the low and nearly constant fading rate in case of the mordanted samples.

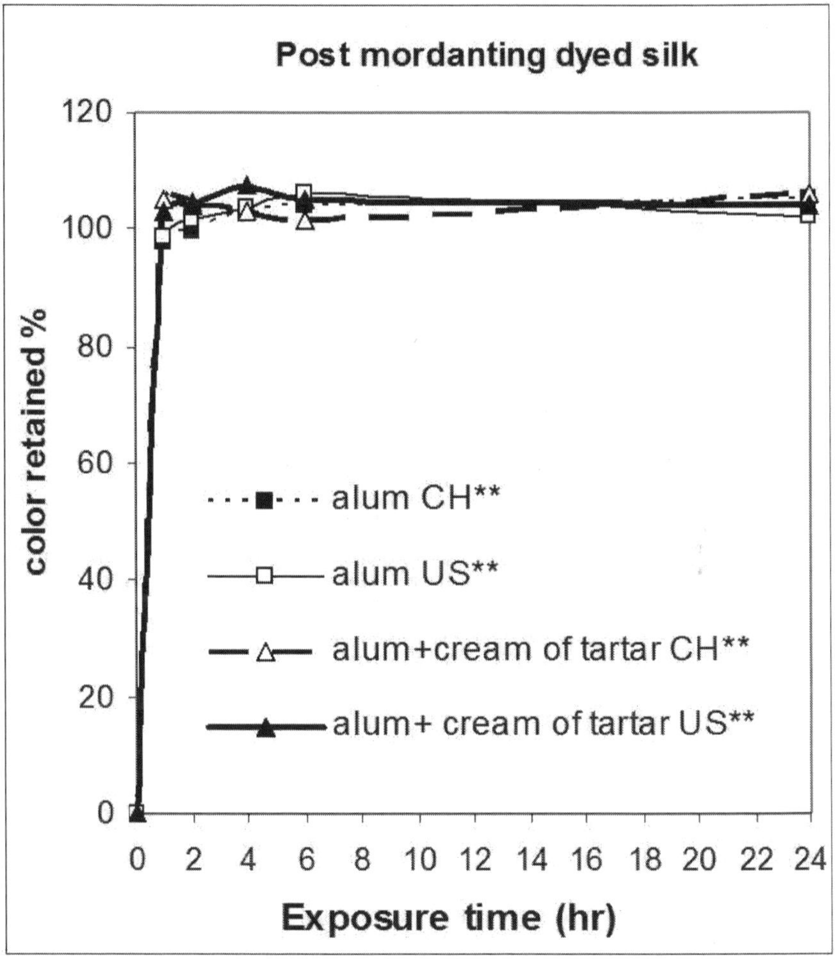

FIGURE 18. Effect of Xenon arc lamp exposure time on the color retained of the post mordanted dyed silk samples

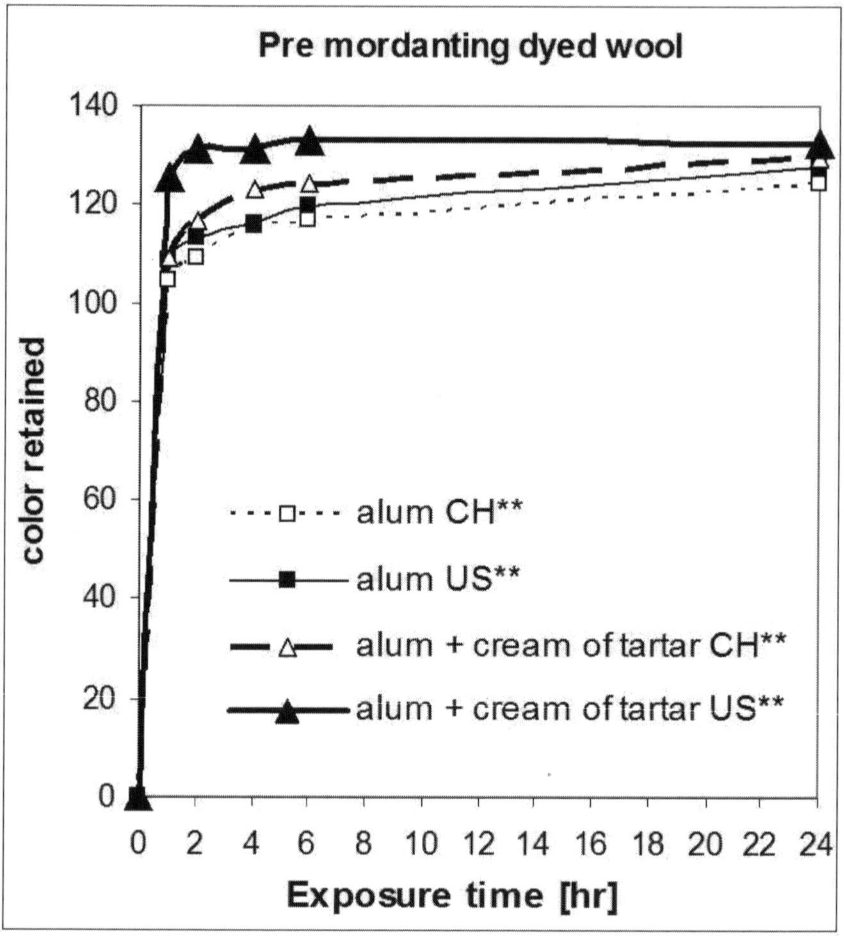

FIGURE 19. Effect of Xenon arc lamp exposure time on the color retained of the pre-mordanted dyed wool samples

4. CONCLUSIONS

With the demand for more environmentally friendly methods and increasing productivity, the newer ultrasonic energy in assisting the extraction of Osage orange natural dye have been developed, over the conventional heating extraction methods, possessing a shorter extraction times and much higher dye absorbance and color strength at lower temperature. Water-acetone co-solvent, and ultrasonic have been found to be the suitable alternatives to the conventional water heating method. The maximum color yield of dye is

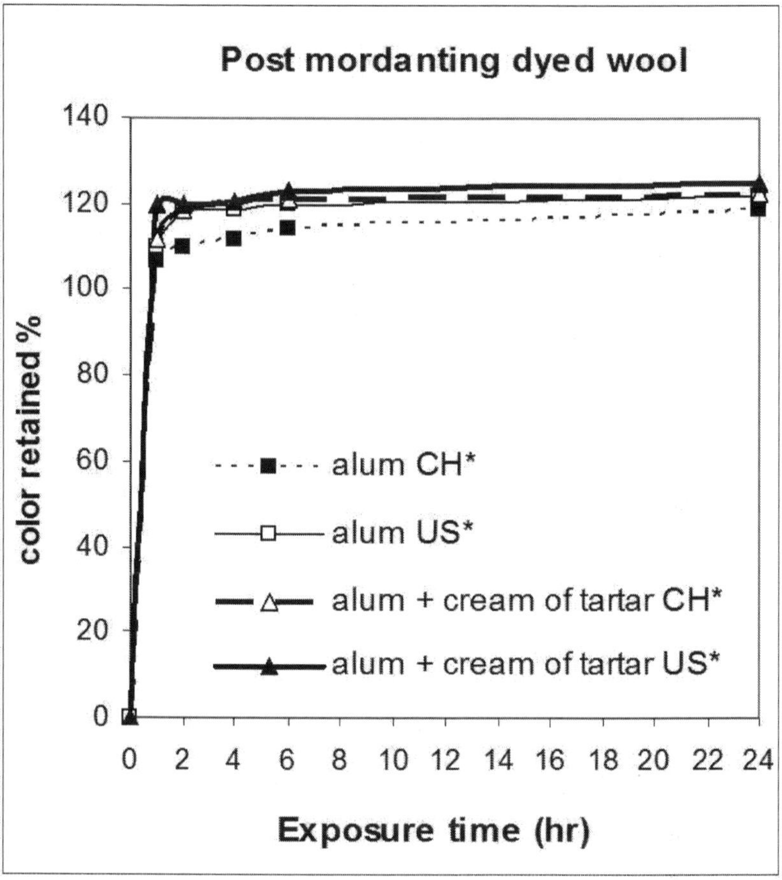

FIGURE 20. Effect of Xenon arc lamp exposure time on the color retained of the post mordanted dyed wool samples

dependent on solvent polarity. Solvation of dye molecules probably occurs via dipole-dipole interactions in non-hydrogen- bond donating solvents, whereas in hydrogen-bond donating solvents the phenomenon is more hydrogen bonding in nature. The dye uptake depends on (The Gilbert- Rideal's and Donnan theories) depending on the coulombic interaction between the anionic groups (OH) in fact: O^- of Osage orange extract and the protonated amino groups of wool fibers.

This research demonstrated the standardization dyeing parameters of Osage orange natural yellow extract on protein based fabrics; silk and wool. The color strength increases with the increase in dyeing temperature and time due to the fiber swelling which enhanced the dye diffusion. The effect of dyeing time was conducted to reveal the effect of power ultrasonic on the de-aggregation of dye molecules in the dye bath. The decline in the dye-ability may be attributed to the hydrolytic decomposition of the extract molecules under the influence of sonic energy during prolonged dyeing. Osage orange is a water-soluble dye containing hydroxyl groups that interacts ionic-ally with the protonated terminal amino groups of silk and wool fibers at acidic pH 5 via ion exchange reaction. The lower dye-ability at pH greater than 5 may be attributed to the enhanced desorption of the dye as its ionic bond is getting decreased. Improvement in the dyeing color strength and fastness properties reflects the higher extent of dye adsorption, interaction and bridging with the pre-mordanted dyed samples via different conjugated bonds with the mixture of alum and cream of tartar mordant. The low color strength in post-mordanting method is due to the accumulation of the aluminum metal dye complex in form of clusters. Sonic energy succeeded in accelerating the rate of mordanting at lower temperatures rather than the conventional heating technique due to the acoustic cavitation which increases the diffusion of solute inside the intermediate spaces of silk and wool fabrics, facilitating the de-aggregation of Osage orange molecules in the dye bath and thus increases the dye diffusion rate, and its penetration through the fibers. A variety of hues were obtained with respect to the mordant. It was observed from the color fastness data that the extracted dye from *Osage orange* furnished different color hues with very good affinity for silk and wool fabrics in presence of a mixture of alum and cream of tartar mordant. Mordanted dyed samples exhibit better wash and light fastness, with a lowest degree of photo fading relative to the control ones.

5. ACKNOWLEDGEMENTS

Special thanks for the school of chemical and physical science, at Victoria university of Wellington in NZ for their financial support. Also many thanks for King abdul-Aziz University, Faculty of Arts and Designs, Fashion Design department for their kindly scientific support

REFERENCES

1. Gulrajani M L. Introduction to Natural Dyes, Indian Institute of Technology. New Delhi: India;.1992

2. Gulrajani ML. Natural Dyes- Part 1, Present status of Natural Dyes. Colourage 1999;. 46(7): 25-27.

3. Gulrajani M L.Present status of natural dyes. Indian Journal Fiber Textile Research. 2001;. 26: 191–201

4. Siva R. Status of natural dyes and dye-yielding plants in India. Current Science 2007; 92 (7): 1-9

5. Hameed BH & El-Khaiary MI. Journal of Hazardous Materials. 2008; 154(1-3): 639-648

6. Georgiou D, Melidis PAivasidis A & Gimouho-poulos K. Dyes and pigments 2002; 52 (2): 69-78.

7. Tsui LS,.Roy WR &.Gole MA. Coloration Technology 2003; 119 (1): 14-8.

8. Araceli R, Juan G, Gabriel O & Maria M. Journal of Hazardous Material 2009;172(2-3): 1311-1320.

9. Uddin MDT, Akhtarul Islam S, Mahmoud Rukanuzzaman MD. Journal of Hazardous Material 2009; 164 (1): 53-60

10. Singh R V. Healthy hues. Down to Earth 2002; 11: 25–31.

11. Tsui LS,.Roy WR &.Gole MA. Coloration Technology 2003; 119 (1): 14-8.

12. Araceli R, Juan G, Gabriel O & Maria M. Journal of Hazardous Material 2009;172(2-3): 1311-1320.

13. Uddin MDT, Akhtarul Islam S, Mahmoud Rukanuzzaman MD. Journal of Hazardous Material 2009; 164 (1): 53-60.

14. Borland VS. Natural resources: animal and vegetable fibers for the 21st century American. Textile. Industry 2000; 29: 66–70

15. Patel B H, Agarwal B J, & Patel H M. Novel padding technique for dyeing Babool dye on cotton. Colourage 2003; January: 21–26.

16. Deo H T & Desai B K. Dyeing of cotton and jute with tea as a natural dye. Coloration Technology 1999; 115(7):224- 227.

17. Aydin A H & Zeki T. 'The sorption behaviors between natural dyes and wool fiber. International Journal of Chemistry 2003; (2): 85-91.

18. Cox P & Leslie P. Osage orange in: Texas Trees, A Friendly Guide. 1999, Cox, P. and Leslie, P. Eds., Corona Publishing Company, San Antonio, TX :147–149.

19. Martins Teixeiraa D, Teixeira da Costa C. Novel methods to extract flavanones and xanthones from the root bark of Maclura pomifera. Journal of Chromatography A 2005; 1062: 175–181.

20. Esra Kupeli, Ilkay Orhan, Gulnur Toker, & Erdem Yesilada. Anti-inflammatory and antinociceptive potential of Maclura pomifera (Rafin.) Schneider fruit extracts and its major isoflavonoids, scandenone and auriculasin. Journal of Ethno pharmacology 2006; 107 (2): 169–174.

21. Tian Li, Jack Blount W & Richard Dixon A. Phenylpropanoid glycosyltransferases from osage orange (Maclura pomifera) fruit. FEBS Letters 2006; 580: 6915–6920.

22. Padma Vankar S, Shanker R, Mahanta D & Tiwari SC Ecofriendly sonicator dyeing of cotton with Rubia cordifolia Linn. Using biomordant. Dyes and Pigments. 2006; 76 (1): 207-212.

23. Mansour HF & Heffernan S. Environmental aspects on dyeing silk fabric with sticta coronata lichen using ultrasonic energy and mild mordants. Clean Technologies and Environmental Policy. 2010; DOI: 10.1007/s10098-010-0296-2.

24. Mansour HF. Environment and energy efficient dyeing of woollen fabric with sticta coronata. Clean Technologies and Environmental Policy. 2009; DOI: 10.1007/s10098-009-0267-7

25. Vajnhandl S, Majcen Le Marechal A. Ultrasound in textile dyeing and the decolouration and mineralisation of textile dyes. Dyes and Pigments 2005; 65 (2):89-101

26. De Santis D. & Moresi M. Production of alizarin extracts from Rubia tinctorum and assessment of their dyeing properties. Industrial Crops and Products 2007; 26 (2): 151–162.

27. Onal A. Extraction of Dyestuff from Onion (Allium Cepa L.) and its application in the dyeing of wool, Feathered, Leather and Cotton. Turkish Journal of Chemistry 1996; 20: 201.

28. Kubelka P. New contribution to the optics of intensity light-scattering materials-Part I. The Journal of Optical Society of America. 1948; 38 (5): 448-451.

29. AATCC. Technical Manual 2000; 75, Research Triangle Park: AATCC.

30. Yoshizumi K & Crews PC. Characteristics of fading of wool cloth dyed with selected natural dyestuffs on the basis of solar radiant energy. Dyes and Pigments 2003; 58 (3): 197–204

31. Crews PC. The fading rates of some natural dyes. Studies in Conservation 1987; 32 (2): 65-72.

32. Giles CH. &.McKay RB. The light fastness of Dyes. Textile Research Journal 1963; 33 (7): 528-77.

33. Muhammad R A, Ahmed SA, and Muhammad K. Solvent effect on the spectral properties of Neutral Red. Chemistry Central Journal 2008; 2:.2-19

34. Oliveira CS, Bronco KP, Baptista MS & Indig GL. Solvent and concentration effects on the visible spectra of tri-para-dialkylamino-substituted triarylmethane dyes in liquid solutions. Spectochim Acta, Part A 2002; 58 {13)2971-2982.

35. Bevilaqua T, Goncalves TF, Venturini CG & Machado VG. Solute-solvent and solvent-solvent interactions in the preferential solvation of 4-[4 (dimethylamino)styryl]-1-methylpyridinium iodide in 24 binary solvent mixtures. Spectochim Acta, Part A 2006; 65 (3): 535-542.

36. Bruce RL, Broadwood NV & King DG. Kinetics of Wool Dyeing with Acid Dyes. Textile Research Journal. 2000; 70 (6): 525-531.

37. Bechtold T, Amalid MA. & Rita Mussak, A M. Reuse of ash-tree (Fraxinus excelsior L.) bark as natural dyes for textile dyeing: process conditions and process stability. Coloration Technology 2007; 123 (4): 271–279.

38. Kongkachuichaya P,Shitangkoonb,A, & Chinwongamorna N. Studies on Dyeing of Silk Yarn with Lac Dye:Effects of Mordants and Dyeing Conditions. Science Asia 2002; 28: 161-166.

39. Mansour HF. Ultrasonic Efficiency on the Photo Fading Of Madder Dyed Silk Using Egg Albumen and Aluminium Ions Chelating. 2009 6th International Textile Conference, NRC, Egypt.

40. Maiwa 2008. http://www.maiwa.com/stores/supply/mordants.html

41. Kamel MM, El-Shishtawy RM, Youssef BM &. Mashaly H. Ultrasonic assisted dyeing. IV. Dyeing of cationised cotton with lac natural dye. Dyes and Pigments 2007; 73 (3): 279-284.

42. Bechtold T, Turcanu A, Ganglberger E., & Geissler S.Natural dyes in modern textile dyehouses — how to combine experiences of two centuries to meet the demands of the future?. Journal of Cleaner Production 2003; 11(5): 499-509.

43. Walter Gardner M. Notes on mordants applicable in wool dyeing. Journal of the Society of Dyers and Colourists 2008; 6 (2): 37 – 40.

44. Gulrajani ML. Natural Dyes- Part 1, Present status of Natural Dyes. Colourage 1999; 46 (7): 25-27.

45. Nadiger GS, Kaushal Sharma P, & Probhaker J. Screening of Natural Dyes. Colourag 2004; 51, annual :130-137.

46. Feng XX, Zhang LL, Chen JY, & Zhang JC. New insights into solar UV protective properties of natural dyes. Journal of Cleaner Production 2007; 15 (4): 366–372.

47. Jjaz A, Bhatti AN, Shahid A, Asghar Jamal M, Safda M & Abbas M. Influence of gamma radiation on the colour strength and fastness properties of fabric using turmeric(Curcuma longa L.) asnaturaldye. Radiation Physics and Chemistry 2010; 79 (5): 622–625.

48. Gupta D, Gulrajani M L, & Kumari S. Light fastness of naturally occurring anthraquinone dyes on nylon. Coloration Technology 2004; 120 (5): 205-212.

49. Clementi C, Nowik W., Romani A, Cibin F & Favaro G. A spectrometric and chromatographic approach to the study of ageing of madder (Rubia tinctorum L.) dyestuff on wool. Analytica Chimica Acta. 2007; 596 (1): 46-54

50. Cristea D & Vilarem G. Improving light fastness of natural dyes on cotton yarn. Dyes and Pigments; 70: 238–245.

51. Mansour HF. Ultrasonic Efficiency on the Photo Fading Of Madder Dyed Silk Using Egg Albumen and Aluminium Ions Chelating. 2009 6th International Textile Conference, NRC, Egypt.

Chapter 3

Impact of Powdered Activated Carbon And Anion Exchange Resin on Photocatalytic Treatment of Textile Wastewater

Preethi Grace Theva Neethi Dhas, Holger Gulyas, Ralf Otterpohl

Institute of Wastewater Management and Water Protection, Hamburg University of Technology, Hamburg, Germany

ABSTRACT

In order to clarify the impact of activated carbon and anion exchange resin on photocatalytic oxidation (PCO) of textile industry wastewater, TiO_2-based PCO was investigated with aqueous solutions containing the reactive dye Reactive Blue 4 (RB4) and with a textile dye house effluent in the absence and in the presence of powdered activated carbon (PAC) and the anion exchange resin Lewatit MP 500. Addition of Lewatit improved RB4 removal to a larger extent than PAC addition. Contrasting to chloride and sulfate, hydrogen carbonate clearly inhibited PCO of RB4. However, the depression of dye removal by hydrogen carbonate was minute in the presence of Lewatit although the hydrogen carbonate concentration was not markedly decreased. Unfortunately, the beneficial effect of Lewatit addition on PCO nearly disappeared when the Lewatit/TiO_2mixture was reused three times. This was probably caused by oxidative damage of the resin. Color removal from the real dye house effluent during PCO was improved by Lewatit, but not by PAC. Contrastingly, PAC addition increased TOC removal by PCO from the real wastewater to some extent, while Lewatit had no impact. Sorbent addition does not lead to an acceptable area demand for solar PCO of the dye house effluent.

Keywords: Activated carbon, Anion Exchange, Hybrid Process, Photocatalytic oxidation, Textile industry Effluent

1. INTRODUCTION

In 2011, the worldwide production of textile amounted to about 25,000 tons cotton fibers, nearly 50,000 tons synthetic fibers, about 100 tons protein fibers and nearly 4500 tons cellulose fibers [1] . Textile industry causes large quantities of wastewaters as the production of one ton textile is accompanied by the generation of 50 to 250 m^3 effluent [1] . Wastewaters are generated predominantly in the following processes of textile production: slashing/sizing, desizing, scouring, bleaching, mercerising, dyeing, printing, and finishing [1] . Characteristics of textile industry effluents are very different and depend on the kind of the manufactured textiles. Besides organic substances, the wastewaters contain also inorganic compounds such as sulfate (which has been determined in high concentrations) [1] , chloride, carbonate, phosphate and heavy metals [2] . Presence of different heavy metals at concentrations ≤ 100 mg/L can be caused by metallized dyes. Cr can originate from acid dyes, Cu from direct dyes, and traces of Hg are present in different raw materials of textile processing [2] . Also organic dyes can exert toxic effects [3] , and anaerobic metabolites of azo dyes represent potential carcinogens [4] . Thus, dyeing, printing and finishing were reported to generate wastewaters with high toxicity [5] . Moreover, raw textile wastewaters after dyeing and finishing often are characterised by intense color [5] . Mondal [6] identified dyes as those constituents of textile wastewater causing the largest difficulties in treatment. The concentration of dyes in textile industry effluents varies greatly with the type of dye used and the kind of dyeing process. Ghaly et al. [1] give 10 to 250 mg/L as a range of dye concentrations in dye house effluents.

Although aerobic biological treatment is generally feasible to remove organics from textile industry effluents [2] , BOD/COD ratios of textile effluents are too low in many cases to operate biological treatment stages successfully [7] . Color removal by this process is ranked from low [2] [8] to insignificant—especially when azo dyes are used [5] . Partial color removal in biological treatment is mainly attributed to precipitation of insoluble dyes and biosorption of soluble dyes [2] . Toxicity of most of the commercial dyes causes problems in biological treatment of textile wastewater such as sludge bulking, sludge rising and inefficient floc formation [5] .

As alternatives to biological treatment, adsorption, ion exchange, membrane processes and oxidative processes are feasible to remove acid, reactive, direct, mordant, metal complex as well as basic dyes. Coagulation, membrane processes and chemical oxidation processes are able to remove disperse, azoic, vat and sulphur dyes [2] . This listing shows that oxidation processes are versatile with respect to dye removal from textile wastewaters. However, chlorine and hydrogen peroxide have been evaluated as unsatisfactory oxidants in comparison to ozone for color removal from the effluent of an MBR treating textile wastewater [9] . Ozonation of biologically pretreated textile wastewater has been shown an efficient color removal technology [10] . However, Robinson et al. [11] mention cost as one of the major drawbacks with ozonation. Also advanced oxidation processes (AOPs) were tested for the treatment of textile dyeing effluents and found to be superior over coagulation [12] .

Heterogeneous photocatalysis using the industrial mass product TiO_2 was recognized as an AOP offering an important advantage over other existing oxidative technologies: It does not require the use of hazardous oxidative substances such as hydrogen peroxide or ozone. Moreover, TiO_2 (which is ideally not consumed because it is used as a catalyst) is non-toxic, chemically stable as well as highly photocatalytically active and inexpensive [13] . Additionally, heterogeneous photocatalytic oxidation (PCO) can be powered by the sun, although there are several challenges to bring solar photocatalysis to technical application [14] .

PCO with TiO_2 as catalyst has been numerously investigated with simulated textile wastewaters which were prepared by dissolving different dyes in water. These investigations were subject of some review articles, e.g. by Ahmed et al. [15] , Akpan and Hameed [3] and Khataee and Kasiri [16] . Rauf et al. [4] reviewed the photocatalytic degradation of azo dyes in the presence of TiO_2doped with transition metals. Alahiane et al. [17] , Chebli et al. [18] , and Kodom et al. [19] have tested PCO of dye solutions with TiO_2 supported on carriers. Solar PCO of the azo dye Acid Orange 7 resulted in considerable color and TOC removal even in North Germany in October [20] . It was also found that the addition of activated carbon to solutions of dyes was beneficial for PCO [21] [22] .

A less extent of research has been executed on PCO of real textile industry wastewaters, e.g. by Chun and Yizhong [23] , Gebrati et al. [24] , Gümüş and Akbal [25] , Hussein [26] , and Pekakis et al. [27] . In those experiments, specific energy consumption by UV lamps was between 270 and 4000 Wh/L. In spite of this high power input, removal of COD and sometimes also of color was not complete.

Also solar PCO experiments were conducted with real textile wastewaters [26] [28] [29] . Although Hussein [26] optimistically concluded from his study that "heterogeneous photocatalytic processes under natural weathering conditions, in the presence of solar radiation show a promising degradation", other studies indicate high area demand for solar PCO of textile wastewaters [28] [30] . The most costly factor in solar PCO is the land requirement [28] .

For making solar photocatalytic oxidation of textile wastewaters a viable process, enhancement of its efficiency is needed. As dyes were recently shown to be well adsorbed to anion exchange resins [31] [32] and also to anion exchange membranes [33] , in this study TiO_2-based PCO of model wastewaters containing Reactive Blue 4 (RB4) and also of a real wastewater was investigated in the presence and in the absence of the anion exchange resin Lewatit MP 500. It was evaluated whether the hybrid process of PCO and anion exchange causes increased PCO efficiency. This hybrid process was compared to another photocatalytic hybrid process: PCO/ powdered activated carbon (PAC). In the experiments with model wastewaters, also the impact of salts on the different PCO processes was investigated.

2. EXPERIMENTAL PROCEDURES

2.1. Textile Dye House Effluent

A textile dye house effluent was transported for experiments from an Indian textile dyeing company. Typical characteristics of the wastewater as determined by the company in India were 1200 mg/L COD, 180 mg/L BOD_5, a low turbidity, a light pink color, and lack of odor. The wastewater was reported to be alkaline (pH 11.7) with a total alkalinity of about 300 mg $CaCO_3$/L and rather saline with a total dissolved solids concentration of around 5300 mg/L and an electric conductivity of 8.5 mmho/cm. The predominant part of the salts was represented by chloride (2000 g/L) and sulfate (1300 mg/L). After shipment, the samples were stored in the refrigerator at 4°C until further analysis and PCO experiments.

2.2. Model Wastewaters

Different model wastewaters were prepared by dissolving 100 mg/L Reactive Blue 4 (RB4) dye (Sigma-Aldrich Taufkirchen, Germany; see Figure 1 for structure) in the following liquids: deionised water, deionised water containing 0.084 mol/L NaCl and 0.010 mol/L Na_2SO_4, and tap water also additionally containing 0.084 mol/L NaCl and 0.010 mol/L Na_2SO_4. NaCl and Na_2SO_4 were purchased in analytical grade from Merck (Darmstadt, Germany).

2.3. Adsorption to Sorbents

The following two sorbents were tested: first, PAC (analytical grade, Merck Eurolab, Darmstadt, Germany, article no. 102186), which exhibits a low ash content (<1%), a BET surface of 775 m^2/g and an average particle size of about 60 μm, and second, Lewatit Monoplus MP 500 (Bayer AG, Leverkusen, Germany) in the chloride form, a strongly basic, macroporous anion exchange resin with beads of uniform size (0.62 mm) based on a styrene-divinylbenzene copolymer. The anion exchange resin was not further prepared by any washing with hydroxide solution, i.e. it was used in the original Cl form. The sorbents were added to a solution of 100 mg/L RB4 in tap water spiked with 0.01 mol/L Na_2SO_4 and 0.084 mol/L NaCl yielding sorbent concentrations of 100, 300, 500 and 1000 mg/L. These suspensions were agitated on a rotational shaker at 150 rounds per minute and 20°C for 72 hours. The supernatants were filtered over membrane filters (Magna Nylon, pore width 0.45 μm, Carl Roth GmbH, Karlsruhe, Germany) and the filtrates analyzed for RB4 concentrations.

2.4. Photolysis and PCO of model Wastewaters

Batches of one liter model wastewater without (photolysis) and with 0.1 g/L TiO_2 photocatalyst (Aeroxide "P25", Evonik Industries AG, Hanau-Wolfgang, Germany) were stirred (magnetic stirrer bar length 7 cm; stirrer

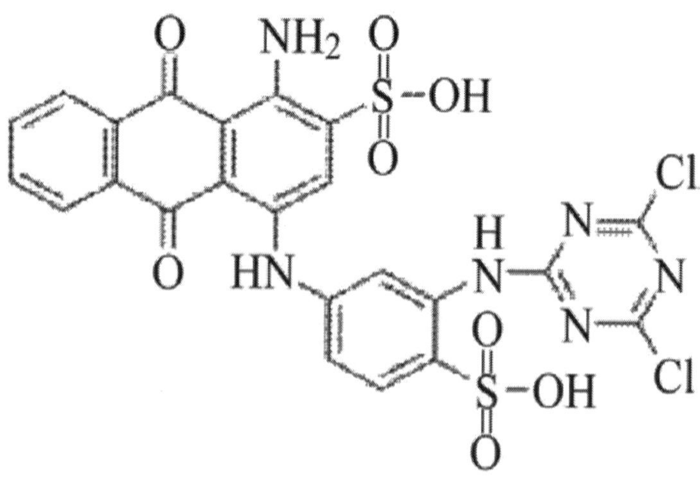

Figure 1. Structure of reactive blue 4.

speed 300 rounds per minute) in slim 2 L beakers (inner diameter 10.8 cm) in the absence and presence of co-adsorbents (0.1 g/L Merck PAC or 0.05 g/L or 0.1 g/L Lewatit anion exchange resin) for at least 42 hours without UV irradiation in order to establish adsorption equilibrium. The solutions (photolysis) and suspensions (PCO) were then UV-A irradiated with face tanners (Philips HB 172; emission maximum at 351 nm) under stirring. The UV lamps were installed at a distance of 20 cm above the liquid surface resulting in a UV intensity of 25 W/m^2 measured with a Pyranometer CM6B (Kipp & Zonen, Delft, Netherlands). During UV irradiation, the temperatures of the reactors were between 21°C and 25°C.

Evaporation loss during UV irradiation was determined by weighing the reactors before and after each sampling and replenished by deionized water before each sampling. Samples of 50 mL were taken at different time intervals, transferred to 100 mL bottles and allowed to settle overnight. The supernatants were centrifuged (Hettich EBA 3 S, 4000 rounds per minute) and centrifugation supernatants were filtered through folded paper filters prior to TOC and photometric analysis.

2.5. Photocatalyst and Sorbents Reuse

A series of experiments were conducted with the hydrogen carbonate-containing model wastewater (100 mg/L RB4 dye in tap water spiked with 0.084 mol/L NaCl and 0.010 mol/L Na_2SO_4) after preparing four different suspensions with the following solids in the model wastewater: first, 0.1 g/L TiO_2, second, 0.05 g/L Lewatit resin, third, 0.1 g/L TiO_2 mixed with 0.1 g/L PAC and fourth, 0.1 g/L TiO_2 mixed with 0.05 g/L Lewatit resin. The suspensions were UV-irradiated for 48 h and the solids were separated from the liquid phase by centrifugation (30 min at 4000 rounds per minute in a Heraeus Multifuge X3, Thermo Scientific) after irradiation was accomplished and reused four times in further PCO cycles with fresh model wastewater. For this purpose, the centrifuged solids were subsequently mixed with 1 L fresh model wastewater and the suspensions irradiated within the next cycle. Samples were collected before addition of catalyst and/or adsorbent and after each 48 h UV irradiation cycle subsequent to separation of the solids from the liquid phase. Irradiated samples were membrane-filtered (0.45 µm pore size) prior to analyses. All samples were analysed for TOC and dye concentration.

2.6. PCO of Real Textile Wastewater

PCO experiments with the real wastewater were performed with an increased amount of catalyst (1 g/L TiO_2) without and with co-adsorbent (0.5 g/L PAC or

0.5 g/L Lewatit) due to the higher TOC concentration in comparison to the model wastewaters. The pH in the reactors was measured once at the beginning and after completion of irradiation. The other conditions were similar to experiments with model wastewaters. Samples were analysed for TOC and color. Experiments with TOC analysis were done in duplicates.

2.7. Analysis

TOC was measured according to the German standard method [34] using a TOC analyser multi-N/C 3000 (Analytik Jena, Jena, Germany). Total inorganic carbon (TIC) and total nitrogen were analysed with the same instrument. Color was determined with a double beam photometer (V-550, Jasco GmbH, Germany) using plastic cuvettes for the model wastewater and quartz glass cuvettes (Suprasil, Hellma GmbH, Müllheim, Germany) for the real wastewater. RB4 concentrations were analysed photometrically by means of calibration standards with known RB4 concentrations. The spectrum of each model wastewater sample and calibration solution was recorded from 800 to 400 nm. The height of the absorption maximum at 599 nm was corrected by subtracting the height of a valley-to-valley baseline from the height of the absorbance maximum. This absorbance difference was linearly related to the dye concentration. In case of the real textile dye effluent, no absorption maximum was observed in the visible range. Absorbance at 420 nm was therefore used as surrogate for color of the wastewater. A WTW SenTix41 probe connected to a Multi 3401 pH meter (WTW, Weilheim, Germany) was used for pH recording.

3. RESULTS AND DISCUSSION

3.1. Adsorption of RB4 from Model Wastewaters

Pure PAC and Lewatit (without being accompanied by TiO_2) adsorbed RB4 from the model wastewater in a tap water matrix additionally containing inorganic salts (Figure 2). The mechanism of RB4 removal by anion ex change can be explained by the two sulfonic acid functional groups contained in the molecule (seeFigure 1) which are converted to anionic sulfonate groups by dissociation of protons. Nevertheless, Lewatit turned out to be a less efficient adsorbent for RB4 than PAC (see Figure 2). Karcher et al. [31] have investigated the sorption of another reactive dye, Reactive Black 5 (2 g/L), which contains four sulfonate groups contrasting to only two in RB4, to a different anion exchange resin (weak basic MP62) the concentration of which was 2 g/L. They found an increased removal of the reactive dye from the aqueous phase when increasing pH from 4.5 (40% removal) to 7.5 (>50% removal). Increase of pH above 8 resulted in even higher removal efficiencies of Reactive Black 5 by the anion exchange resin MP62 [31] . The results in the study of Karcher et al. [31] were better than the 20% removal of RB4 from 100 mg/L solutions with a Lewatit concentration of 100 mg/L at pH 7.4 as shown in Figure 2.

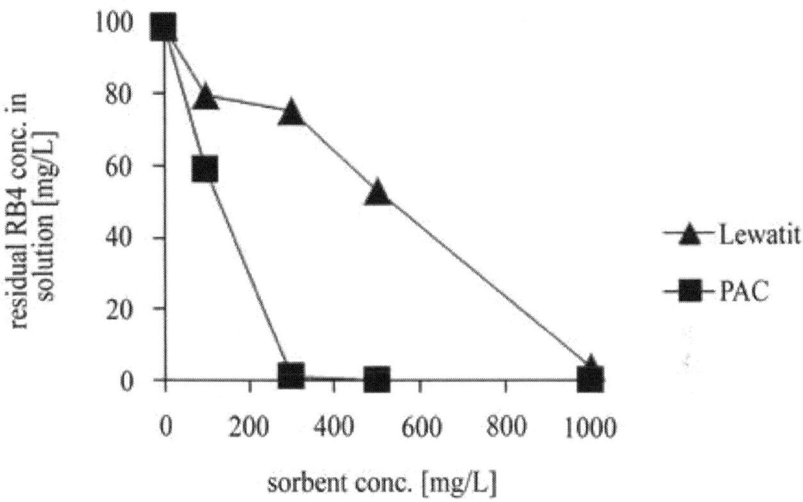

Figure 2. Adsorptive RB4 removal by Lewatit and PAC record- ed at 20°C with a 100 mg/L RB4 solution in tap water spiked with 0.084 mol NaCl/L and 0.01 mol Na_2SO_4/L (pH 7.4).

Figure 2 shows that with 50 mg/L PAC, a reduction in RB4 concentration of about 20% was achieved. This was contrasting to a 73% RB4 removal during stirring of the same model wastewater with 100 mg/L TiO_2 and 50 mg/L PAC even when considering a 14.7% RB4 removal by adsorption to TiO_2 (Table 1). There was also a substantial decrease of RB4 concentration by nearly 70% subsequent to stirring the suspensions without UV when PAC was replaced by Lewatit (Table 1). This removal by adsorption to a mixture of TiO_2 and Lewatit also cannot be explained by summing up the 14.7% adsorption to TiO_2 (Table 1) and about 10% adsorption to pure Lewatit (Figure 2).

When a higher concentration (100 mg/L) of Lewatit was added to the TiO_2 suspension in the RB4 matrix in deionized water spiked with NaCl and Na_2SO_4, there was nearly no color detectable even without UV irradiation (Table 1). Overall, the Lewatit/TiO_2 mixture is a promising adsorbent for RB4. As another advantage of adding Lewatit to PCO, better settling of the TiO_2 nanoparticles along with the resin was observed.

It was interesting that the inorganic matrix of the model wastewater did not markedly affect the adsorption of RB4 to TiO_2 and to TiO_2/Lewatit mixture, while adsorption to the TiO_2/PAC was larger in the saline matrices than in the deionized water matrix.

3.2. Photolysis and PCO of Model Wastewaters

Figure 3 shows concentrations of RB4 and TOC in different RB4 solutions in different matrices during UV-A illumination. RB4 was only slightly susceptible to photolysis and there was no clear effect of the inorganic wastewater matrix on photolysis (Figure 3(a)). No mineralization took place during UV-A irradiation (Figure 3(b)).

Stirring TiO_2 with the different model wastewaters for <42 h without UV-A irradiation resulted in about 15% removal of RB4 by adsorption to the photocatalyst (Table 1). UV-A illumination of the suspensions led to a further decrease in RB4 concentration (Figure 4(a)). Chloride and sulfate addition to deionized water did not markedly decrease PCO efficiency (compare curves "salts/DI water" and "DI water" in Figure 4(a)). It can thus be concluded that chloride and sulfate were not massively depressing RB4 PCO efficiency, although inorganic salts have been discussed by many authors to deteriorate the PCO process [35] - [39] . Alahiane et al. [17] who investigated the impact of different inorganic salts on photocatalytic oxidation of Reactive Yellow 145 by

Table 1. Removal of RB4 by adsorption from the tested model wastewaters to TiO_2 (100 mg/L) and to mixtures of TiO_2 with PAC or Lewatit during the stirring period of >42 h without UV prior to PCO (DI water = deionized water).

	RB4 removal in the dark [%]		
	DI water	Salts/DI water	Salts/tap water
TiO_2	15.4	14.2	14.7
TiO_2 + 50 mg/L PAC	36.2	79.0	73.0
TiO_2 + 50 mg/L Lewatit	68.6	74.8	68.8
TiO_2 + 100 mg/L Lewatit	-	98.8	-

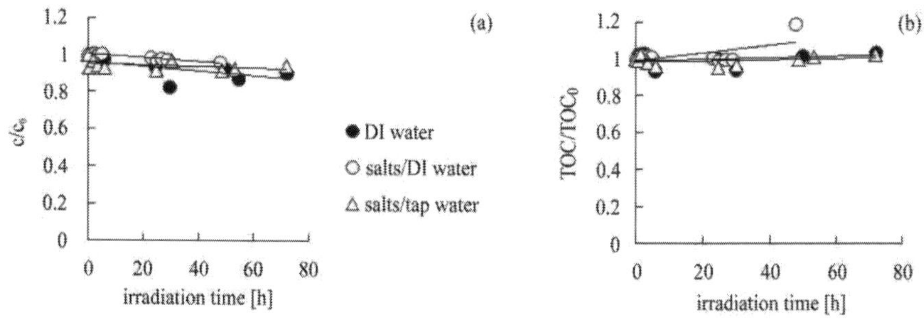

Figure 3. Normalized RB4 (a) and TOC concentrations (b) of 100 mg/L RB4 solutions in deionized water, in a deionized water matrix containing additionally 0.084 mol/L NaCl and 0.01 mol/L Na_2SO_4 and in a tap water matrix containing additionally 0.084 mol/L NaCl and 0.01 mol/L Na_2SO_4during photolysis.

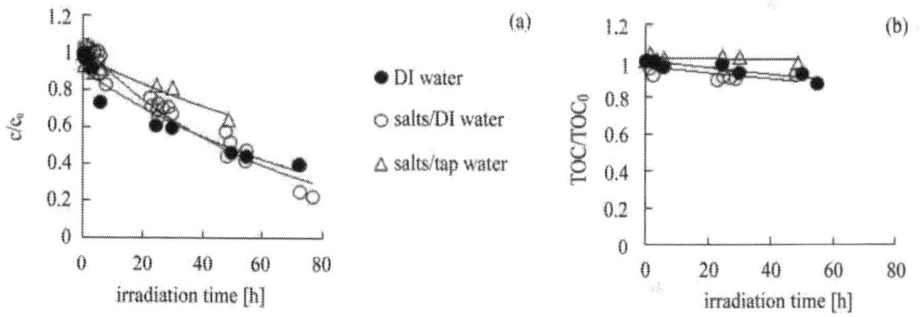

Figure 4. Normalized RB4 (a) and TOC concentrations (b) of 100 mg/L RB4 solutions in deionized water, in a deionized water matrix containing additionally 0.084 mol/L NaCl and 0.01 mol/L Na_2SO_4 and in tap water containing additionally 0.084 mol/L NaCl and 0.01 mol/L Na_2SO_4 during PCO without adsorbents.

TiO_2-coated natural and synthetic non-woven filters found even an increase of PCO efficiency by chloride and sulfate.

In order to simulate the real textile dye house effluent which contains considerable amounts of hydrogen carbonate as indicated by the total alkalinity of about 300 mg $CaCO_3$/L, RB4 solutions were also prepared in tap water spiked with NaCl and Na_2SO_4 in similar concentrations as in the real textile wastewater. RB4 removal by PCO in the tap water matrix with salts was slightly less efficient than in the deionized water matrix (open triangles in Figure 4(a)) which can be explained by the radical scavenging effect of hydrogen carbonate contained in the tap water which exhibited a TIC concentration of 22 mg/L (equivalent to 112 mg/L HCO_3^-). Alkalinity has been recently shown to have an inhibitory effect on PCO [30] .

The pH of the TiO_2 suspensions in the deionised water matrix with NaCl and Na_2SO_4 (which did not contain hydrogen carbonate) was 4.8, while it was 7.4 in the tap water matrix with NaCl and Na_2SO_4. Therefore, also the pH might have influenced the PCO efficiency. Kodom et al. [19] ascertained that photocatalysis of Reactive Black 5 with TiO_2 coated on paper showed the highest efficiency at pH 3, while the activity was less at pH 7. This was explained by predominance of negatively charged surface groups on the photocatalyst surface at pH 7 resulting in decreased adsorbability of the dye's sulfonate groups. Similarly, increased COD removal from a real textile industry effluent as well as increased decolorisation of an aqueous Remazol Red 133 solution were achieved by photocatalytic oxidation at pH 3 in comparison to pH 9.5 [25] .

Demineralization of RB4 solutions by PCO was negligible (Figure 4(b)). In the presence of the radical scavenger hydrogen carbonate, nearly no mineralization took place within 50 h UV irradiation in the presence of TiO_2 (open triangles in Figure 4(b)).

In the presence of PAC, PCO of RB4 was clearly inhibited in the tap water matrix. According to [17] , chloride and sulfate did not decrease the efficiency of decolorisation by PCO in the presence of PAC (Figure 5(a)). However, mineralisation in the PCO/PAC hybrid process was depressed by salts (open circles in Figure 5(b)). A comparison of Figure 4(a) to Figure 5(a) indicates that the addition of PAC promotes RB4 removal percentage by PCO. Also the mineralization of RB4 solutions in deionized water was enhanced by PAC addition (com- pare filled circles in Figure 4(b) to those in Figure 5(b)), but not in the salt-containing matrices. However, also for the salt-containing model wastewaters, the PAC/PCO hybrid process was beneficial, as PAC addition resulted in 70% - 80% adsorptive RB4 removal (Table 1).

The addition of the anion exchange resin Lewatit MP 500 to TiO_2 was very promising for color removal from all tested model wastewaters, because besides good adsorptive removal of RB4 by TiO_2/Lewatit mixtures (Table 1), it also efficiently promoted the photocatalytic RB4 removal in all matrices (Figure 6(a)). It was surprising that PCO was even more efficient in the salt-containing model wastewaters. Obviously, addition of Lewatit was able to depress the radical-scavenging effect of hydrogen carbonate on decolorization (open triangles in Figure 6(a)) but not on mineralization (open triangles in Figure 6(b)). However, the larger slope of the curve for photocatalytic oxidation in the presence of Lewatit (Figure 6(a)) in comparison to the curve for mere PCO (Figure 4(a)) cannot be attributed to adsorptive removal of hydrogen carbonate by Lewatit, because Lewatit addition decreased TIC only marginally from 22 mg/L to 18.7 mg/L measured at the end of the PCO/Lewatit hybrid experiment (data not shown).

3.3 Multiple Reuse of Photocatalyst and Sorbents

Long-term dye removal capacity of Lewatit and different photocatalytic procedures can be visualised when the solids are reused several times. Figure 7 shows residual dye (a) and TOC (b) concentrations (related to the concentrations of the model wastewater before addition of the solids) after 48 h UV irradiation of the salt-contain- ing model wastewater (100 mg/L RB4, 0.084 mol/L NaCl and 0.01 mol/L Na_2SO_4 in a tap water matrix) in the presence of TiO_2, TiO_2/PAC and TiO_2/Lewatit which had been recycled by centrifugation subsequent to each irradiation cycle and added to a fresh batch of model wastewater for another illumination cycle. The results from photometric measurements (Figure 7(a)) and from TOC analyses (Figure 7(b)) correlated to some extent.

In Figure 7(a), it can be seen that the PCO/Lewatit hybrid process shows a similar RB4 removal (42%) as the PCO/PAC process when the solids were used for the first time (solids reuse cycle 0). PCO (21 % RB4 removal) was less efficient than the hybrid processes, because the latter ones included RB4 adsorption to PAC or Lewatit. The major difference to the experiments described in Section 3.2 was the lack of pre-agitation of the suspensions prior to UV illumination. From Table 1and Figures 4(a)-6(a), it can be concluded that the investigated processes with stirring of suspensions prior to UV irradiation showed the following rank order for RB4 removal from the model wastewater in the tap water matrix spiked with chloride and sulfate: PCO/Lewatit (98% RB4 removal) > PCO/PAC (81% RB4 removal) > PCO (45% RB4 removal). So, the RB4 removal in the experiments with pre- agitation prior to UV irradiation was definitely larger.

In the PCO/PAC hybrid process (crosses in Figure 7(a) and Figure 7(b)), dye and TOC removal efficiency were markedly decreased from about 40% (when using fresh solids) in subsequent cycles of TiO_2/PAC reuse reaching the same low level as PCO without PAC (open circles) in reuse cycles 2, 3 and 4. This indicates that the PAC was not regenerated during the preselected 48 h PCO period and that much longer UV irradiation times would be needed for the photocatalytic PAC regeneration. In a recent study about the PCO/PAC hybrid process for TOC removal from biologically pretreated greywater [40] , the TOC subsequent to UV irradiation also increased from one TiO_2/PAC reuse cycle to the next. However, the initial TOC of the pretreated grey water was only half as much as that of the model wastewater used in this study, and the PCO/PAC hybrid process for grey water treatment was more efficient than mere PCO even after several reuse cycles.

The Lewatit suspension without TiO_2 (open triangles in Figure 7) was not exposed to UV light, but only stirred for the same period as the other suspensions were irradiated for. When the Lewatit concentration was 50

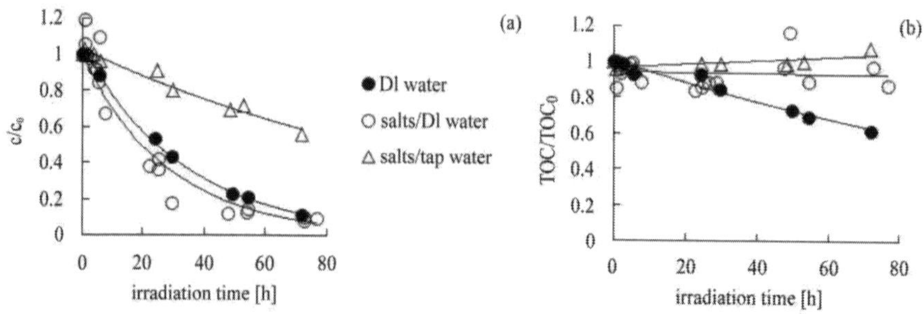

Figure 5. Normalized RB4 (a) and TOC concentrations (b) of 100 mg/L RB4 solutions in deionized water, in a deionized water matrix containing additionally 0.084 mol/L NaCl and 0.01 mol/L Na_2SO_4 and in tap water containing additionally 0.084 mol/L NaCl and 0.01 mol/L Na_2SO_4 during PCO in the presence of 100 mg/L PAC.

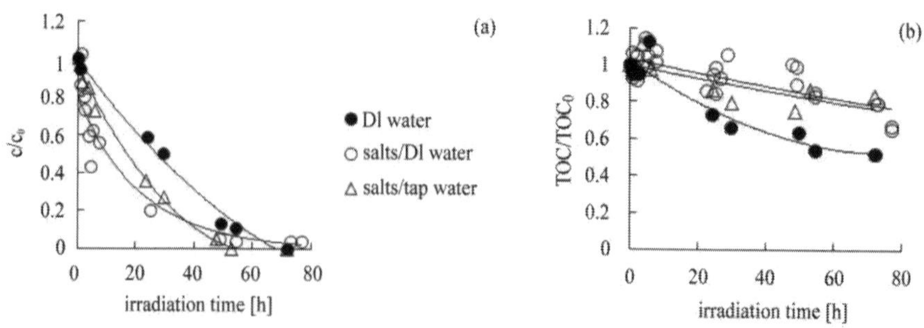

Figure 6. Normalized RB4 (a) and TOC concentrations (b) of 100 mg/L RB4 solutions in deionized water, in a deionized water matrix containing additionally 0.084 mol/L NaCl and 0.01 mol/L Na_2SO_4 and in tap water containing additionally 0.084 mol/L NaCl and 0.01 mol/L Na_2SO_4 during PCO in the presence of 50mg/L Lewatit.

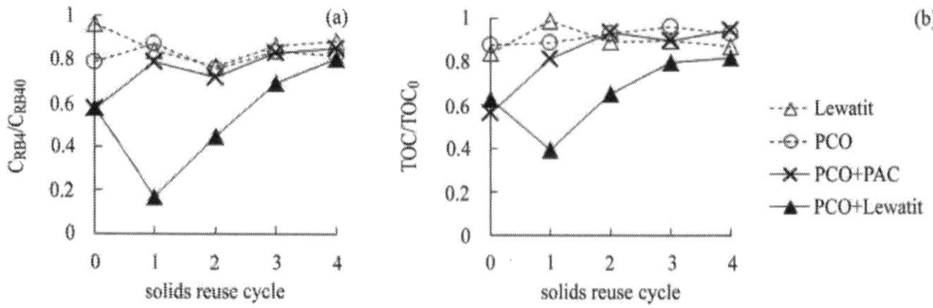

Figure 7. Normalized RB4 (a) and TOC (b) concentrations of salt-containing model waste- water (100 mg/L RB4, 0.084 mol/L NaCl and 0.01 mol/L Na_2SO_4 in tap water) subsequent to 48 h treatment (PCO, PCO/PAC, PCO/Lewatit and stirring with 50 mg/L Lewatit without UV) within 5 cycles of use of the different solids.

mg/L, the anion exchange resin was obviously not completely loaded with RB4, because there was a reduction of dye and TOC concentration (approximately 10% - 20%) even in Lewatit reuse cycles 1 to 4 (open triangles in Figure 7) indicating that the adsorption capacity of the Lewatit was not yet completely expired even in the fourth reuse cycle. However, the fresh Lewatit (which had not been pretreated prior to use) showed only a marginal dye removal (figure 7(a)). This is according to poor RB4 adsorption to 50 mg/L Lewatit as displayed in Figure 2. However, it is in contradiction to very good RB4 adsorption from the same model wastewater to Lewatit when TiO_2 was added as shown inTable 1. In the suspension containing the Lewatit/TiO_2 mixture either some photocatalytic removal of RB4 was caused in the pre-irradiation period by lighting in the laboratory, or the addition of TiO_2improved RB4 adsorption to Lewatit by effects which currently cannot be explained.

The Lewatit/PCO hybrid process with fresh solids (filled triangles in Figure 7(a) and Figure 7(b)) caused a dye removal of 42%. This was much lower than after 48 h UV irradiation in the experiment with pre-agitation in the dark displayed in Figure 3(c). A possible explanation for this contradiction is that adsorption to Lewatit was less efficient in the experiments on TiO_2/Lewatit reuse, because there was no stirring of the suspensions prior to the PCO period without UV irradiation for at least 42 hours. In contrast to the cycle with fresh Lewatit, nearly all dye was removed in the PCO/Lewatit hybrid process when the Lewatit/TiO_2 was reused for the first time. Unfortunately, in subsequent reuse cycles of the TiO_2/Lewatit mixture, the dye removal efficiency was massively reduced. In reuse cycle 4 it was reduced to the efficiency of the PCO process without co-adsorbent. This reduction in efficiency can be explained by progressing oxidative damage of the organic anion exchange resin within several solids reuse cycles or by insufficient photocatalytic regeneration of the

Lewatit. This indicates that the Lewatit/PCO hybrid process—in spite of large decoloration efficiency when using TiO_2/Lewatit as fresh material or reusing it only once—is not recommendable because fresh resin has to be added at least after two to three reuse cycles.

Improvement of TOC removal from solids reuse cycle 0 to 1 in the PCO/Lewatit hybrid process (filled triangles in Figure 7(b)) can be explained as follows: In "solids reuse cycle 0", Lewatit was used without prior washing. A fishy smell indicated its content of amine from production of the anion exchange resin. This amount of amine probably caused some additional TOC. In the first reuse cycle, there was no more amine present on the Lewatit because it was washed to the liquid phase in the prior cycle. This might have resulted in a higher TOC removal in reuse cycle no. 1. However, this is no explanation for improved RB4 removal by the reused TiO_2/ Lewatit mixture during UV irradiation (filled triangles in Figure 7(a)).

3.4. PCO of Real Textile Dye House Effluent

The textile effluent had to be transported from India to Germany and hence the chemical characteristics of the effluent might have changed during transportation. It took four days for the samples to reach destination. The parameters of the wastewater measured in the laboratory in Hamburg are shown in Table 2. While the fresh ori- ginal wastewater was reported to be odourless, it exhibited some fishy smell when it was used for the experi- ments. Additionally it was observed that the wastewater caused eye irritation during experimentation. Fishy smell and eye irritation may be referred to the generation of anilines from residual azo dyes during anaerobic storage of the wastewater sample. Both phenomena were no longer realized when the PCO experiments were accom- plished. Also the color of the wastewater was no longer light pink as described by the company, but brown when it arrived at Hamburg (Table 2). This might also be a consequence of azo dye decay. The TOC concentration of 765 mg/L classifies the wastewater as suitable for being treated by an advanced oxidation process. However, it is in a relatively high range for this class of processes.

An absorbance maximum in the visible range could not be established from the spectrum recorded with the real textile wastewater (Figure 8). Only a small part of the major absorption band of the wastewater was in the visible range > 400 nm. However, this was responsible for the brown color of the effluent. Decoloration was thus quantified by means of the absorbance at 420 nm.

Experiments on adsorption of the wastewater organics to PAC, Lewatit and TiO_2 (data not shown) indicated the possibility of applying higher concentrations of photocatalyst (1 g/L) and co-adsorbent (0.5 g/L) for the real wastewater still resulting in detectable color and TOC concentrations after contact of the wastewater with the solids. Because of residual organics being present after adsorption, color and TOC removal could be quantified in subsequent photocatalysis.

The pH was reduced from 12.4 in the beginning to around 9 at the end of the PCO experiments. This decrease in pH may be caused by degradation of anilines and is in accordance with the disappearance of the fishy smell after accomplishing PCO experiments.

Figure 9(a) shows that TiO_2 and the TiO_2/PAC mixture did not adsorb the colored substances from the real wastewater, whereas the TiO_2/Lewatit mixture even without UV irradiation well removed dye molecules absorbing light at 420 nm. The mechanism behind the dye sorption cannot be explained, since the textile effluent probably contains many different dyes in small concentrations. It might be assumed that most of the residual dyes were reactive dyes, because those were used for dyeing cotton according to company information. At least, the efficient decoloration by Lewatit indicates that the color of the wastewater was mainly due to anionic substances. Probably, the residual dyes also contained sulfonate groups. Photocatalytic oxidation itself and also thePCO/PAC hybrid process were not that efficient for color removal. However, PCO could add some color re moval subsequent to Lewatit adsorption. Overall, the adsorption to Lewatit with additional PCO resulted in a color removal of about 80%.

PCO treatment without the addition of sorbents did not show efficient mineralisation (Figure 9(b)). The contact of the real wastewater with Lewatit/TiO_2 without UV illumination contributed only to a small TOC removal (about 8%). Photocatalytic TOC removal was not increased to a large extent by the addition of 0.5 g/L of the co- adsorbent Lewatit. The Lewatit/PCO hybrid process led to a TOC removal of only 22% to 34%.

Table 2. Real textile dye house effluent parameters.

Parameters	Values
pH	12.4
Color	Brown
Odour	Fishy
TOC (mg/L)	765
Total nitrogen (mg/L)	27.8
TIC (mg/L)	170

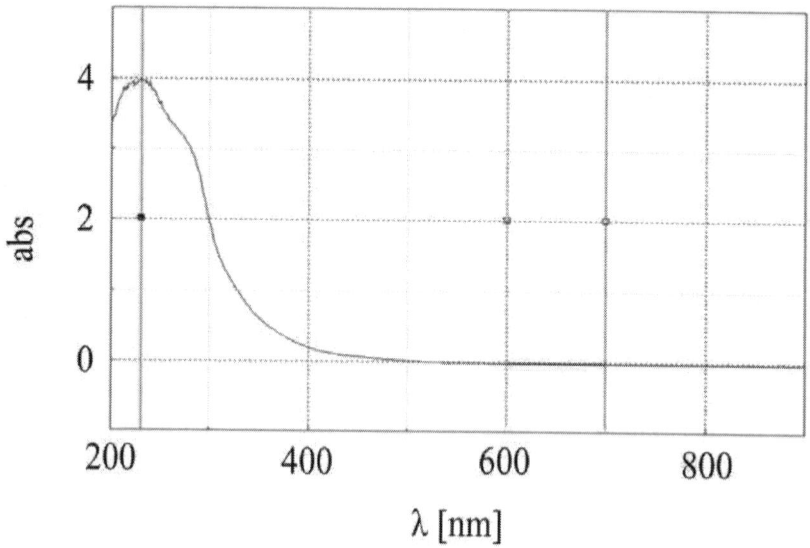

Figure 8. Absorption spectrum of the real textile dye house effluent.

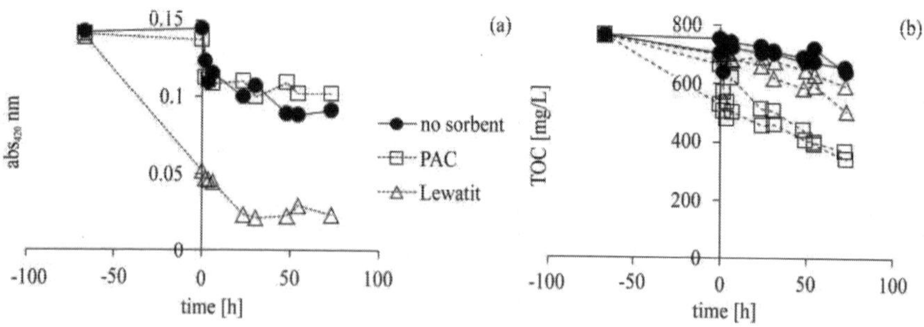

Figure 9. Removal of absorbance at 420 nm (a) and TOC (b) from real textile wastewater by PCO with 1 g/L TiO_2 without and with additional sorbents (0.5 g/L PAC or 0.5 g/L Lewatit); negative time axis represents stirring of the suspensions without UV irradiation.

Adsorption to PAC/TiO_2 without UV removed 13% - 30% of TOC with high data variation in duplicate experiments. At the end of the irradiation period, 52% - 55% of TOC were removed by the PCO/PAC hybrid process. The PCO process in the presence of PAC showed a higher TOC removal than in the presence of Lewatit or in the absence of adsorbents, although the addition of PAC is expected to absorb UV photons making the process less efficient. A possible explanation of this contradiction might be the adsorption of wastewater constituents inhibiting the PCO process to PAC but not to Lewatit. On the other hand, Lewatit well adsorbed the colored substances.

An important aspect for process implementation is the energy requirement for PCO. The UV dose, Q_{UV}, introduced into the reactor can be calculated by Equation (1).

$$Q_{UV} = \frac{I \cdot A \cdot t}{V_L} \tag{1}$$

where I is the UV intensity in W/m^2, A is the liquid surface area in m^2, t is the irradiation time in h and V_L is the volume of wastewater in the reactor in L. Considering the reactor content of 1 L, an inner diameter of the reactor of 10.8 cm, a UV intensity of 25 W/m^2, and an irradiation time of 70 h, a UV dose of 16.5 Wh/L was introduced into the reactor.

Either power consumption will be high when UV lamps are used for the process or land area demand will be very large when the photocatalytic process is designed to be powered by the sun. India is a sub-tropical country receiving adequate solar UV radiation around 10 months in a year [41] . As an estimate for the lowest insolation in the year at Tiruppur (in November), 4.3 $kWh/m^2/d$ can be selected from the respective Gaisma website [42] . Considering the required UV dose of 16.5 kWh/m^3 and a minimum solar UV irradiation (5% of global irradiation) of 0.215 $kWh/m^2/day$ as calculated from the value given above, the area required to treat one m^3 of textile dye house effluent within one day would be about 77 m^2. This would result in a wastewater layer with a depth as small as 1.3 cm. Without covering the reactor with UV-translucent material, this shallow layer will be completely evaporated within one day preventing wastewater reuse. It has to be pointed out that this large area would only result in a TOC removal by the PCO/PAC hybrid process of not much more than 50% as can be seen in Figure 9(b). As also other investigations [28] [30] indicated large area demand for solar PCO of textile wastewaters, large volumes of textile wastewater are not recommended to be treated by this solar process.

4. CONCLUSION

Although color was removed by PCO from aqueous RB4 solutions used as simulated wastewater, the mineralisation of RB4 by this process was poor. The addition of chloride and sulfate concentrations found in real textile wastewaters did not deteriorate the efficiency of photocatalytic RB4 degradation. However, the radical scavenger hydrogen carbonate depressed efficiency of PCO as shown by experiments with model wastewater prepared in a tap water matrix. Color as well as TOC removal by PCO was intensified when sorbents (powdered activated carbon or the anion exchange resin Lewatit) were added to the PCO process. It turned out that RB4 was more efficiently adsorbed to PAC and Lewatit when TiO_2 was present. This could not be explained by the presented data. Although the anion exchange resin was more efficient in enhancing dye removal by PCO than PAC and although it prevented the process from inhibition by the radical-scavenging hydrogen carbonate, hybridisation of the PCO process with Lewatit anion exchange was no longer efficient, when the Lewatit/TiO_2 mixture was reused for several UV irradiation cycles. This might be caused by incomplete photocatalytic regeneration or by oxidative damage of the anion exchange resin. Because of good decolorisation of the real textile industry wastewater by the PCO/Lewatit hybrid process and improved TOC removal by the PCO/PAC hybrid process, it was recommended to use anion exchange as a first stage for color removal and PCO in the presence of PAC for textile wastewater treatment rather than PCO alone. From the experimental data, an area demand for solar PCO hybridised with PAC addition of nearly 80 m^2 was calculated for the treatment of 1 m^3 of the investigated textile industry effluent within one day. Unfortunately, this will result in a residual TOC concentration of still 370 mg/L. This huge requirement of area for treatment in a highly industrialised area makes solar PCO unsuitable. On the other hand, it is essential that textile dye house effluents are toxic, which eventually justifies a high investment for their treatment to achieve zero liquid discharge. A feasible option would be the use of UV lamps instead of sunlight. Research on PCO has to be advanced in order to minimize the area demand for the solar PCO process.

ACKNOWLEDGEMENTS

The authors gratefully acknowledge the support of Susanne Eggers for TOC analyses. They also express their gratitude to the German Federal Ministry of Education and Research (BMBF) for having funded the Master studies of Preethi Grace Theva Neethi Dhas within the International Postgraduate Studies on Water Technologies (IPSWAT) program, without which this research would not have been possible.

REFERENCES

1. Ghaly, A.E., Ananthashankar, R., Alhattab, M. and Ramakrishnan, V.V. (2014) Production, Characterization and Treatment of Textile Effluents: A Critical Review. Chemical Engineering & Process

Technology, 5, 182. http://omicsonline.org/open-access/production-characterization-and-treatment-of-textile-effluents-a-critical-review-2157-7048.1000182.pdf

2. Correia, V.M., Stephenson, T. and Judd, S.J. (1994) Characterisation of Textile Wastewaters—A Review. Environ- mental Technology, 15, 917-929.

3. Akpan, U.G. and Hameed, B.H. (2009) Parameters Affecting the Photocatalytic Degradation of Dyes Using TiO_2- Based Photocatalysts: A Review. Journal of Hazardous Materials, 170, 520-529.

4. Rauf, M.A., Meetani, M.A. and Hisaindee, S. (2011) An Overview on the Photocatalytic Degradation of Azo Dyes in the Presence of TiO_2 Doped with Selective Transition Metals. Desalination, 276, 13-27.

5. Pang, Y.L. and Abdullah, A.Z. (2013) Current Status of Textile Industry Wastewater Management and Research Progress in Malaysia: A Review. Clean Soil Air Water, 41, 751-764.

6. Mondal, S. (2008) Methods of Dye Removal from Dye House Effluent—An Overview. Environmental Engineering Science, 25, 383-396.

7. Mountassir, Y., Benyaich, A., Rezrazi, M., Berçot, L. and Gebrati, L. (2013) Wastewater Effluent Characteristics from Moroccan Textile Industry. Water Science & Technology, 67, 2791-2799.

8. Lotito, A.M., De Sanctis, M., Di Iaconi, C. and Bergna, G. (2014) Textile Wastewater Treatment: Aerobic Granular Sludge vs Activated Sludge Systems. Water Research, 54, 337-346.

9. Brik, M., Chamam, B., Schöberl, P., Braun, R. and Fuchs, W. (2004) Effect of Ozone, Chlorine and Hydrogen Peroxide on the Elimination of Colour in Treated Textile Wastewater by MBR. Water Science & Technology, 49, 299-303.

10. Orhon, D., Dulkadiroğlu, H., Doğruel, S., Kabdaşli, I., Sozen S. and Babuna, F.G. (2002) Ozonation Application in Activated Sludge Systems for a Textile Mill Effluent. Water Science & Technology, 45, 305-313.

11. Robinson, T., McMullan, G., Marchant, R. and Nigam, P. (2001) Remediation of Dyes in Textile Effluent: A Critical Review on Current Treatment Technologies with a Proposed Alternative. Bioresource Technology, 77, 247-255.

12. Azbar, N., Yonar, T. and Kestioglu, K. (2004) Comparison of Various Oxidation Processes and Chemical Treatment Methods for COD and Color Removal from a Polyester and Acetate Fiber Dyeing Effluent. Chemosphere, 55, 35-43.

13. McCullagh, C., Skillen, N., Adams, M. and Robertson, P.K.J. (2011) Photocatalytic Reactors for Environmental Re- mediation: A Review. Journal of Chemical Technology and Biotechnology, 86, 1002-1017.

14. Gulyas, H. (2014) Solar Heterogeneous Photocatalytic Oxidation for Water and Wastewater Treatment: Problems and Challenges. Journal of Advanced Chemical Engineering, 4, Article ID: 1000108. http://omicsonline.com/open-access/solar-heterogeneous-photocatalytic-oxidation-for-water-and-wastewater-treatment-problems-and-challenges-2090-4568-4-108.pdf

15. Ahmed, S., Rasul, M.G., Martens, W.N., Brown, R. and Hashib, M.A. (2011) Advances in Heterogeneous Photocata- lytic Degradation of Phenols and Dyes in Wastewater: A Review. Water Air Soil Pollution, 215, 3-29.

16. Khataee, A.R. and Kasiri, M.B. (2010) Photocatalytic Degradation of Organic Dyes in the Presence of Nanostructured Titanioum Dioxide: Influence of the Chemical Structure of Dyes. Journal of Molecular Catalysis A: Chemical, 328, 8- 26.

17. Alahiane, S., Qourzal, S., El Quardi, M., Abaamrane, A. and Assabbane, A. (2014) Factors Influencing the Photocata- lytic Degradation of Reactive Yellow 145 by TiO_2-Coated Non-Woven Fibers. American Journal of Analytical Chemi- stry, 5, 445-454.

18. Chebli, D., Fourcade, F., Brosillon, S., Nacef, S. and Amrane, A. (2010) Supported Photocatalysis as a Pretreatment Prior to Biological Degradation for the Removal of Some Dyes from Aqueous Solutions; Acid Red 183, Biebrich Scarlet, Methyl Red Sodium Salt, Orange II. Journal of Chemical Technology and Biotechnology, 85, 555-563.

19. Kodom, T., Akpénè, D., Tchakala, I., Gnazou, M.E.D.T., Djaney-Boundjou, G. and Bawa, M.L. (2013) TiO_2 PC500 Coated on Non Woven Paper with SiO_2 as a Binder-Assisted Photocatalytic Degradation of Reactive Black 5 in Aqueous Solution. Journal of Water Resource and Protection, 5, 1227-1234.

20. Gulyas, H., Stürmer, R. and Hintze, L. (2001) Experiences with Solar Application of Photocatalytic Oxidation for Dye Removal from a Model Textile Industry Wastewater. In: Brebbia, C.A., Ed., Water Pollution

VI—Modelling, Measur- ing and Prediction, WIT Press, Southampton, 153-165.

21. Jamil, T.S., Ghaly, M.Y., Fathy, N.A., Abd el-halim, T.A. and Österlund, L. (2012) Enhancement of TiO_2 Behavior on Photocatalytic Oxidation of MO Dye Using TiO_2/AC under Visible Irradiation and Sunlight Irradiation. Separation and Purification Technology, 98, 270-279.

22. Sobana, N. and Swaminathan, M. (2007) Combination Effect of ZnO and Activated Carbon for Solar Assisted Photo- catalytic Degradation of Direct Blue 53. Solar Energy Materials Solar Cells, 91, 727-734.

23. Chun, H. and Yizhong, W. (1999) Decolorization and Biodegradability of Photocatalytic Treated Azo Dyes and Wool Textile Wastewater. Chemosphere, 39, 2107-2115.

24. Gebrati, L., Loukili Idrissi, L., Mountassir, Y. and Nejmeddine, A. (2010) Détoxification des effluents d'industrie de textile par photocatalyse. Environmental Technology, 31, 625-632.

25. Gümüş, D. and Akbal, F. (2011) Photocatalytic Degradation of Textile Dye and Wastewater. Water Air & Soil Pollu- tion, 216, 117-124.

26. Hussein, F.H. (2012) Comparison between Solar and Artificial Photocatalytic Decolorization of Textile Industry Waste- water. International Journal of Photoenergy, 2012, Article ID: 793648. http://dx.doi.org/10.1155/2012/793648http://www.hindawi.com/journals/ijp/2012/793648/

27. Pekakis, P.A., Xekoukoulotakis, N.P. and Mantzavinos, D. (2006) Treatment of Textile Dyehouse Wastewater by TiO_2 Photocatalysis. Water Research, 40, 1276-1286.

28. Bousselmi, L., Geissen, S.-U. and Schroeder, H. (2004) Textile Wastewater Treatment and Reuse by Solar Catalysis: Results from a Pilot Plant in Tunesia. Water Science & Technology, 49, 331-337.

29. Vilar, V.J.P., Pinho, L.X., Pintor, A.M.A. and Boaventura, R.A.R. (2011) Treatment of Textile Wastewaters by Solar- Driven Advanced Oxidation Processes. Solar Energy, 85, 1927-1934.

30. Gulyas, H., Jain, H.B., Susanto, A.L., Malekpur, M., Harasiuk, K., Krawczyk, I., Choromanski, P. and Furmanska, M. (2005) Solar Photocatalytic Oxidation of Pretreated Wastewaters: Laboratory Scale Generation of Design Data for Tech- nical-Scale Double-Skin Sheet Reactors. Environmental Technology, 26, 501-514.

31. Karcher, S., Kornmüller, A. and Jekel, M. (2002) Anion Exchange Resins for Removal of Reactive Dyes from Textile Wastewaters. Water Research, 36, 4717-4724.

32. Greluk, M. and Hubicki, Z. (2011) Efficient Removal of Acid Orange 7 Dye from Water Using the Strongly Basic Anion Exchange Resin Amberlite IRA-958. Desalination, 278, 219-226.

33. Liu, C.-H., Wu, J.-S., Chiu, H.-C., Suen, S.-Y. and Chu, K.H. (2007) Removal of Anionic Reactive Dyes from Water Using Anion Exchange Membranes as Adsorbers. Water Research, 41, 1491-1500.

34. Normenausschuss Wasserwesen (1997) Determination of Total Organic Carbon (TOC) (H 3) (in German), DIN EN 1484 (1997-08), German Standard Methods for the Examination of Water, Waste Water and Sludge, DIN Deutsches Institut für Normung e.V., Berlin.

35. Abdullah, M., Low, G. and Matthews, R. (1990) Effects of Common Inorganic Anions on Rates of Photocatalytic Oxidation of Organic Carbon over Illuminated Titanium Dioxide. Journal of Physical Chemistry, 94, 6820-6825.

36. Bekbölet, M., Boyacioglu, Z. and Özkaraova, B. (1998) The Influence of Solution Matrix on the Photocatalytic Re- moval of Color from Natural Waters. Water Science & Technology, 38, 155-162.

37. Calza, P. and Pelizzetti, E. (2001) Photocatalytic Transformation of Organic Compounds in the Presence of Inorganic Ions. Pure Applied Chemistry, 73, 1839-1848.

38. Guillard, C., Puzenat, E., Lachheb, H., Honas, A. and Herrmann, J.-M. (2005) Why Inorganic Salts Decrease the TiO_2 Photocatalytic Efficiency. International Journal of Photoenergy, 7, 1-9.

39. Zhang, W., An, T., Cui, M., Sheng, G. and Fu, J. (2005) Effects of Anions on the Photocatalytic and Photoelectrocata- lytic Degradation of Reactive Dye in a Packed-Bed Reactor. Journal of Chemical Technology and Biotechnology, 80, 223-229.

40. Gulyas, H., Choromanski, P., Muelling, N. and Furmanska, M. (2009) Toward Chemical-Free Reclamation of Biologi- cally Pretreated Greywater: Solar Photocatalytic Oxidation with Powdered Activated Carbon. Journal of Cleaner Pro- duction, 17, 1223-1227.

41. Shanmuga Priya, S., Premalatha, M. and Anantharaman, N. (2008) Solar Photocatalytic Treatment of Phenolic Waste- water—Potential, Challenges and Opportunities. ARPN Journal of Engineering and

Applied Sciences, 3, 36-41.http://www.arpnjournals.com/jeas/research_papers/rp_2008/jeas_1208_141.pdf

42. GAISMA (Without Date) Tiruppur, India—Solar Energy and Surface Meteorology.http://www.gaisma.com/en/location/tiruppur.html

Chapter 4

Evaluation of Economical and Ecological Aspects of Denim Garments Dyeing with Fluorescent Dye

Muhammad Abdur Rashid[1], Md. Delwar Hossain[1], Muhammad Mufidul Islam[2], Md. Nakib-Ul-Hasan[2]

[1]Department of Textile Engineering, Dhaka University of Engineering & Technology (DUET), Gazipur, Bangladesh

[2]Department of Textile Engineering, Northern University Bangladesh, Dhaka, Bangladesh

ABSTRACT

In this study the existing process (3 steps) and proposed process (2 steps) were evaluated during dyeing of denim fabric with fluorescent dye. It compared the effectiveness of these two processes of fluorescent dyeing in the textile industry. Here at first the grey denim fabric was pretreated by scouring and bleaching. Then the fabric was dyed in existing process i.e. through catanizing, dyeing and binding at 10% and 15% shades. Again the pretreated fabric was dyed in proposed process i.e. through catanizing and (dyeing and binding) with same shades. Then the properties of two types of dyed fabric were compared. This study provides a set of experimental results, discussion and comparison between the two processes. The results show that the proposed process has less power and time consumption and more water savings compared to existing process. So the evaluation reflected that the proposed process was much more effective than existing process of denim dyeing with fluorescent dye.

Keywords: Fluorescent Dyestuff; Denim Fabric; Fastness Properties

1. INTRODUCTION

Bangladesh is in 4th position in textile in world survey [1]. In our country mainly woven and knit fabrics are used in garments sector. Dyed fabric or white fabrics are used in garments sector. Their almost all dye used in Bangladesh. Nowadays the application of fluorescent dye for both woven and knitted fabrics are increasing day by day for its brilliance color appearance although the fastness properties are not so good.

There are several very early reports of the usual fluorescent brilliance of certain dyes when applied to natural textile fiber, the most pronounced effect being obtained on silk. The greenish-yellow fluorescent and it's orange to red halogenated derivatives (eosin) were among the dyes but there practical use was limited by inadequate fastness properties. The brilliant bluish red rhodamines were found to exhibit rather better fastness properties on textiles and they are the most important classes of fluorescent dyes for a wide range of application. Research aimed at producing new fluorescent dyes for using in the coloration of textile, coating and plastics has continued in recent years with the objectives of extending the available product shade range enhancing fluorescent intensity and improving other useful application properties such as light fastness and thermal stability. In addition several new applications have emerged which specifically exploit light emission properties of fluorescent dyes [2]. Techniques involving the use of specific fluorescent dyes are of immense importance in biological and medical research providing a useful Lolo in the search for new biologically active compounds and the potential for the development of new diagnostic methods [3]. It is difficult to arrive at a precise definition of what constitutes a fluorescent dye. Florescent dye is a chemical that contains a lot of property. Especially it is renowned for its special photo physical property. A lot of people are trying to implement this dye on textile field. TENCEL? (Lyocell) and other man-made cellulosic fibers using a fluorescent dye molecular probe [4]. Only TENCEL shows a significant difference between dried and never-dried fibers and it is mainly due to the structure of TENCEL. But if it is done on the other method it expects that it may bring an outstanding result. The use of hemi cyanine dyes to bring the effect that is similar to fluorescent effect [5]. The results showed that those dyed acrylic fabrics using hemi-cyanine dyes had obvious fluorescent effect in the spectra range 550 - 750 nm. But the effect that is brought by florescent is not same here. So it is possible to make an experiment by using fluorescent dye in the process reduction method that brings an effect in the spectra range 550 - 750 nm. An improvement in fastness to sublimation; washing and perspiration (both acidic and alkaline) were observed but there was no general improvement in light fastness compared with the parent compounds [6]. The fluorescent azo disperse dyes were applied on polyester fiber and they finally probe that it enhances various fastness property [7]. Another experiment was carried out on wool, silk and nylon with fluorescent acid azo dyes and obtained photo physical properties increased but no fastness property was noticed [8]. The maximum fluorescent reflectance would reach under acid medium dyeing (pH = 5), adding fluorescent brightener can obviously raise the reflectance [9]. At the same time other properties remain constant. So it would be very interesting if it is possible to increase other properties with the reflectance property. The experimental results show that more complete chromatogram of fluorescent acrylic yarn can be obtained with suitable proportion of mixed dyestuff in combination dyeing [10]. The purpose of this research is to reduce the process steps of fluorescent dyeing of denim garments without changing fabric properties including ecological and economical benefit.

2. MATERIALS AND METHODS

2.1. Materials

The investigation has been carried out with 100% cotton 3/1 twill denim fabric. The yarn count was 7 $^{s}/_{1}$ carded yarn, the fabric weight per unit area was 372 g/m^2, EPI 68 and PPI 38 respectively. The fabric was scoured and bleached by caustic soda and hydrogen per oxide in a single bath stage with the standard recipe [11]. Then the denim fabrics were dyed with fluorescent dye in various shade percentage.

2.2. Methods

2.2.1. Pretreatment Process

The denim cotton fabric has been subjected to scouring and bleaching process for removal natural impurities such as wax, pectin, fat, oil, pigment etc from cotton fibers [12]. Here scouring and bleaching was carried out by using suitable recipe (**Table 1**) with maintaining proper time and temperature. After pretreatment the fabric was washed off two times i.e. hot wash and cold wash and then dried. Then the pretreated denim fabric was used for the experimental process. The same pretreated sample has been used for existing process and also for the proposed process.

2.2.2. Dyeing

1) Existing Process The pretreated fabric was treated with 5 gm/l catanizer (Cat MDN-96, Texodyes Company, Bangladesh) at 50˚C temperature for 20 minutes in pH 8.5. Then the dye bath was prepared by 10% and 15% of fluorescent red, fluorescent yellow and fluorescent blue separately and combined in the liquor

ratio 1:10. After catanizing, the fabric was performed cold wash. Then the sample was added to each dye bath and dyeing was commenced 10 minutes at room temperature then raised temperature to 60˚C for 20 minutes. The dyed sample was washed with cold water and treated with 3 gm/l binder for 5 minutes and finally cold wash at room temperature and then dried in oven.

Table 1. Recipe of scouring and bleaching process.

Process parameter	Amount
Wetting agent	1 gm/l
Sequestering agent	1 gm/l
Detergent	1 gm/l
Caustic soda	4 gm/l
Hydrogen per oxide	5 gm/l
Stabilizer	2 gm/l
Temperature	95˚C
Time	1 hour

2) Proposed Process The pretreated fabric was treated with 5 gm/l catanizer (Cat MDN-96, Texodyes Company, Bangladesh) at 50˚C temperature for 20 minutes in pH 8.5. Then the dye bath was prepared by 10% and 15% of fluorescent red, fluorescent yellow and fluorescent blue separately and combined in the liquor ratio 1:10. After catanizing, the fabric was performed cold wash. Then the sample was added to each dye bath and dyeing was commenced 20 minutes at temperature 60˚C with 3 gm/l binder. Finally the dyed fabric was performed cold wash at room temperature and then dried in oven.

2.2.3. Fabric Parameter Test

Treated all fluorescent dyed denim fabric were conditioned in 65% RH and at 20˚C for 24 hours before testing according to ASTM D1776 [13]. Shade matching was performed by data color machine. Dye yield capacity (k/s) of the fabric was tested according to kubel-kumung theory by data color machine. Fabric brightness test was also performed by data color machine. The values for color fastness were rated with a grey scale for color change and color staining according to AATCC test method 61 [14]. SEM was studied using a scanning electron microscope (model-S 3400N, Hitachi, Japan).

3. RESULT AND DISCUSSION

3.1. Color Difference Measurement

Calculation of small color Differences for acceptability CMC (Color Measurement Committee) values for different shade percentages are estimated from the dyed fabrics by existing process & proposed process. The results are shown (**Table 2**).

CIE (Commission Internationale de l'Eclairage) color coordinates include color qualities in terms of L* (lightness and darkness), a* (redness and greenness), b* (yellowness and blueness), c* (chroma) and H (hue) of the fabrics are shown in **Table 2**. The ΔE values of differently dyed fabrics. ΔE value contains the information of color depth, shade, and the hue of a sample. ΔE value is calculated by using the CIE L*, a*, and b* values with the equation $\Delta E = (\Delta L^{*2} + \Delta a^{*2} + \Delta b^{*2})^{1/2}$where ΔL*, Δa*, and Δb* values were the difference between L*, a*, and b* values of a pair of color standard and sample. The larger the ΔE value, the greater will be the color difference between this pair of color standard and sample. In this case the color difference was expected because the CMC value was less than 1.0 after dyeing existing and proposed process. The experiment was carried out under light source of D65 at 10° and the result shows the satisfactory result remarks as "pass".

3.2. Color Fastness to Rubbing

In this project work color fastness to rubbing assessments are done for both fabrics dyed by existing process and proposed process. This test was designed to determine the degree of color which may be transferred from the surface of a colored fabric to a specify test cloth for rubbing which could be dry and wet. The dyed fabric was subjected to test fastness to rubbing. This test result showed in **Table 3**, where good rubbing fastness for 10% and 15% shade of proposed process compared to existing process dyed of denim fabrics. Because the time for migration of dye was more in proposed process. The experiment was carried out according to the ISO105X12 method. **Table 3** shows the test results.

3.3. Color Fastness to Wash

In this experiment color fastness to wash assessments are used for dyed fabric both existing and proposed process. The resistance of a material to change in any of its color characteristics, when subjected to washing is called color fastness to washing. In the test, change in color of the textile and also staining of color on the adjacent fabric were assessed.

Color fastness was usually assessed separately with respect to fading and staining. This test shows (**Figure 1**) good to excellent wash fastness for color changed and staining of the proposed process compared to the existing process. Because the reaction of proposed process has combined than the existing process.

3.4. Power Consumption

The power requirement for proposed process is less than existing process. **Table 4** shows the comparison

Figure 2 indicates that the power requirement in proposed process is less than in existing process which is about 25% .The evaluation was carried out for 100 kg garments.

3.5. Time Consumption

The time requirement for proposed process is less than existing process. The existing process takes place about 1 hour and 15 minutes for completion whereas the proposed process will not take more than 55 minutes for completion the whole process of dyeing. Thus for every batch dyeing the proposed process will save 20 minutes. It will also reduce the production cost of the whole process and adding the economical benefit. In terms of percentage the time saving will be 26.67% which is shown in **Table 5" target="_self"> Table 5**. The whole process was evaluated for 100 kg garments.

Table 2. CMC values of existing process & proposed process dyed fabric.

Shade %	ΔE	ΔL*	Δa*	Δb*	Δc*	ΔH	CMC Value
10%	1.98	0.89	1.59 (-)	0.77 (-)	0.34 (-)	1.74	0.88
15%	1.97	1.00	1.70 (-)	0.11	0.52	1.62	0.86

Table 3. Comparison of color fastness to rubbing of existing and proposed process.

Process	Shade	Test Result		Sample view	
		Dry	Wet	Dry	Wet
Existing Process	10%	3/4	2/3		
	15%	3	2		
Proposed Process	10%	4/5	2/3		
	15%	4/5	3		

Table 4. Power consumption in existing and proposed processes.

Process	Required Power in Unit (Kilo Watt/Hour)	Power saved in %
Existing Process	20	25
Proposed Process	15	

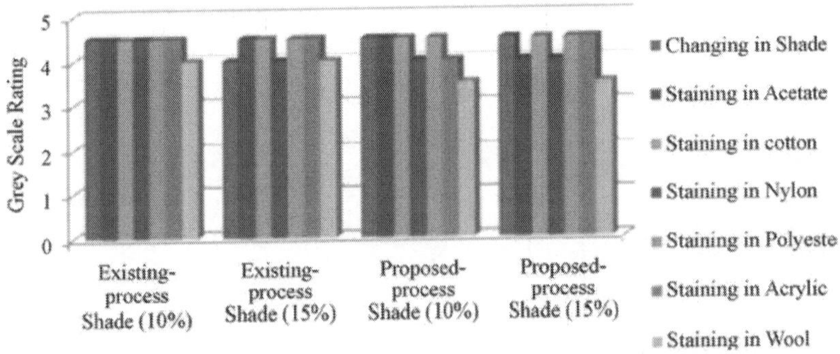

Figure 1. Graphical representations of color fastness to wash for existing and proposed process.

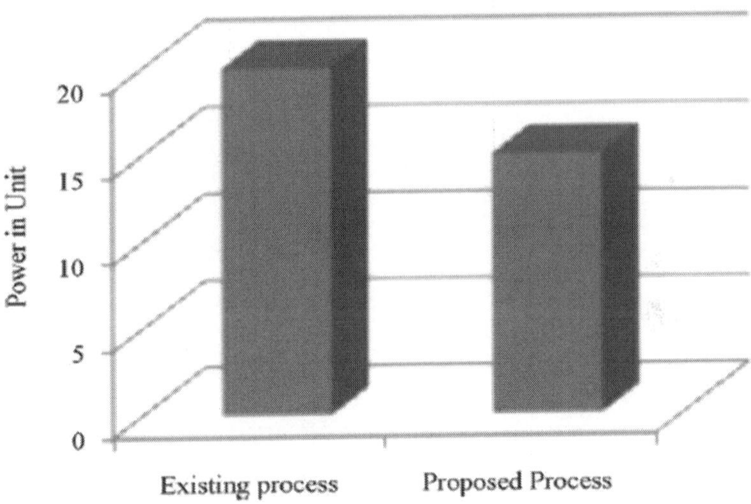

Figure 2. Power requirements in existing and proposed processes.

3.6. Water Consumption

The water requirement for proposed process is less than existing process. **Table 6" target="_self"> Table 6** shows the comparison In proposed process two baths are used whereas in existing process a minimum of three baths are used. This leads to a total one bath saving. Considering material liquor ratio of 1:10; the total amount of water that will be saved is about 20 liters for every kg garments for dyeing. For above result 100 kg garments were undertaken for considering the total result.

3.7. Surface View on Micro Macro Projector

The Microscopic view of grey fabric, dyed fabric (proposed process, exsiting process) are given below (see Figures 5-7):

The Figures (Figures 5-7) show that there is a little bit difference among surface views of grey fabric, fabric dyed by existing process & proposed process. Finally it is observed that the surface views of fabric dyed by proposed process at different shade % are comparatively smoother than any other.

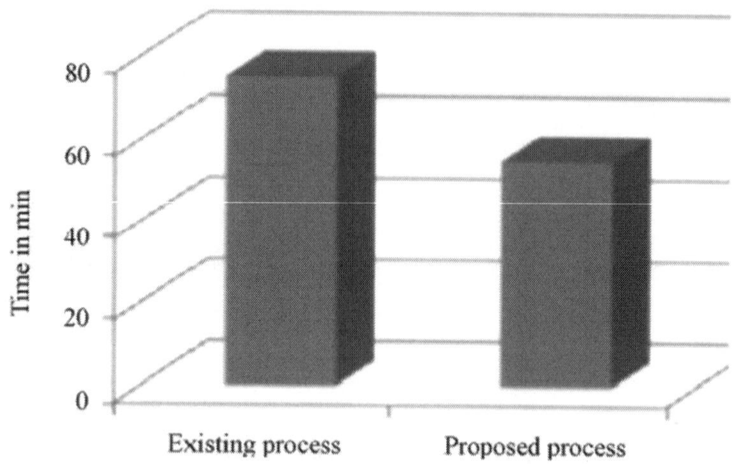

Figure 3. Time requirement for existing and proposed process.

Table 5. Time comparison in existing and proposed processes.

Process	Required Time (min)	Time saved in %
Existing Process	75	26.67
Proposed Process	55	

Table 6. Water requirements for existing and proposed processes.

Process	Required water in Liter	Water saved in %
Existing Process	5981	33.43
Proposed Process	3981	

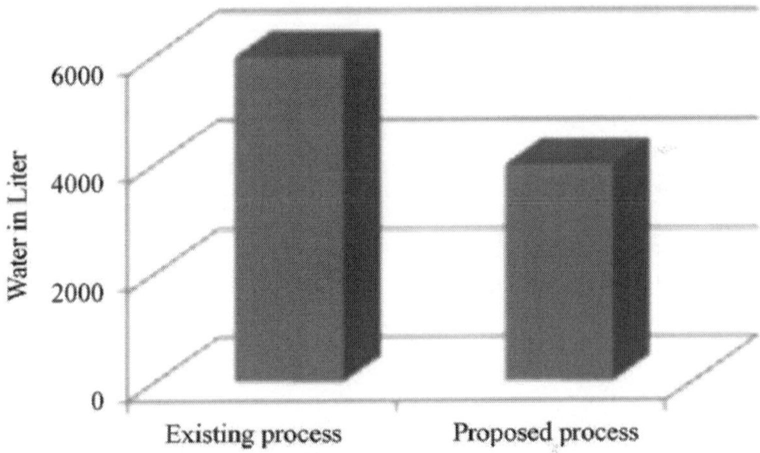

Figure 4. Water requirements in existing and proposed processes.

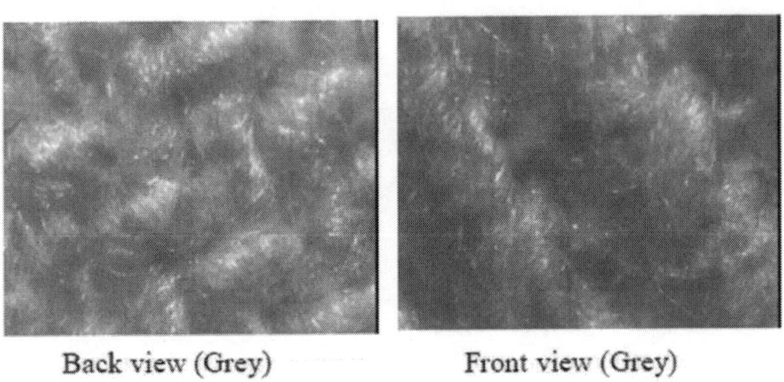

Figure 5. Microscopic view of grey fabric.

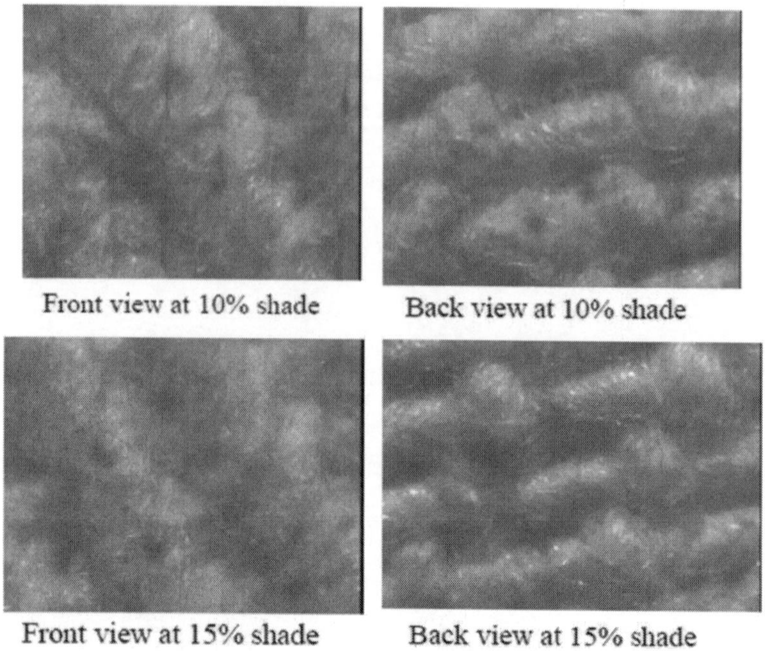

Figure 6. Microscopic view of dyed fabric by existing process.

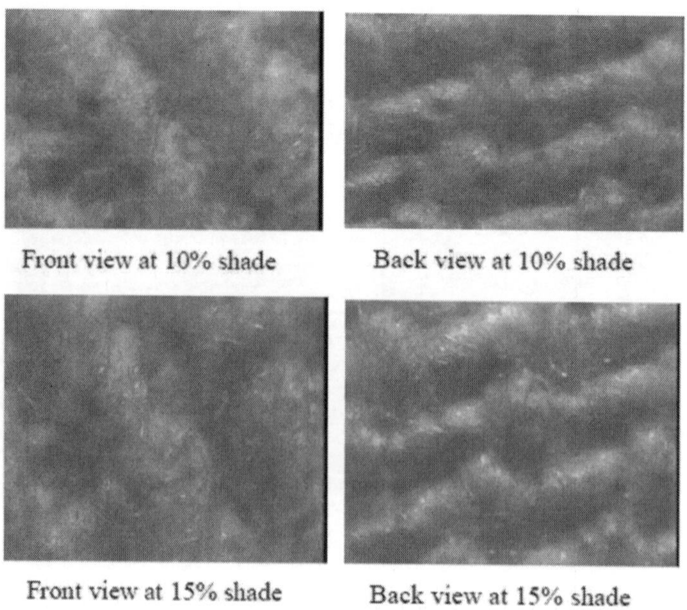

Figure 7. Microscopic view of dyed fabric by proposed process.

4. CONCLUSION

The results show that the properties of the dyed fabric with existing and proposed process remain similar. It was observed that the value of CMC less than 1 i.e. 0.88 and 0.86 of 10% and 15% shaded respectively of existing and proposed process. But the fastness properties of the proposed process were better than the existing process. Another important activity of the research work was to analyze the impact of the proposed process an economical and ecological aspect which shows that proposed process give better result compared to the existing process. It is shown that the proposed process saved about 25% of power consumption, 26.67% of time consumption and 33.43% of water consumption compared to the existing process. The study clearly indicates the economical savings of production cost also scope for a sustainable process development to support the eco-friendly textile processing by releasing high pressure on energy and utility consumption.

REFERENCES

1. Dorpon Textiles Limited, Source: EPB compiled by RDTI. 2011. www.dorpangroup.com/dtextiles/default.aspx

2. T. Gülümser, E. Karagöz and E. P. A Kumbasar, "A Research About Dyeing Of Cotton Fibers With Fluorescent Dyestuffs And Developing Light Fastness," Tekstil Ve Konfeksiyon, 2008.

3. G. Cosa1, K.-S. Focsaneanu, J. R. N. McLean, J. P. McNamee and J. C. Scaiano, "Photophysical Properties of Fluorescent DNA-Dyes Bound to Singleand Double-stranded DNA in Aqueous Buffered Solution," Photochemistry and Photobiology, Vol. 73, No. 6, 2001, pp. 585-599.

4. M. Abu-Rous, K. Varga, T. Bechtold and K. C. Schuster, "A New Method to Visualize and Characterize the Pore Structure of TENCEL� (Lyocell) and Other Man-Made Cellulosic Fibres Using a Fluorescent Dye Molecular Probe," Article First Published Online: 23 JUL 2007.

5. C. X. Qin, R.-C. Tang, B. Chen, D. W. Chen, X. M. Wang and Q. Chen, "Study on the Dyeing Properties of Hemicyanine Dyes. I. Acrylic Fabrics, Fibers and Polymers," Fibers and Polymers, Vol. 11, No. 2, 2010, p. 8.

6. R. M. Christie, K. M. Morgan and M. S. Islam, "Molecular Design and Synthesis of N-Arylsulfonated Coumarin Fluorescent Dyes and Their Application to Textiles," School of Textiles and Design, Scottish

Borders Campus, Heriot-Watt University, Galashiels TD1 3HF, Scotland, 2008.

7. M. A. Satam, R. K. Raut and N. Sekar, "Fluorescent Azo Disperse Dyes from 3-(1, 3-Benzothiazol-2-yl) Naphthalen-2-Ol and Comparison with 2-Naphthol Analogs," Dyes and Pigments, Vol. 96, No. 1, 2013, pp. 92-103.

8. M. A. Satam, R. K. Raut and N. Sekar, "Fluorescent Acid Azo Dyes from 3-(1,3-Benzothiazol-2-yl)Naphthalen-2- Ol and Comparison with 2-Naphthol Analogs," Dyes and Pigments, Vol. 97, No. 1, 2013, pp. 32-42.

9. W.-F. Yang and L.-J. Yang, "Polyamide Dyeing with Fluorescent Dye," Tianjin Polytechnic University, Tianjin, 2004.

10. L. Wang and L. Hu, "Combination Dyeing and One-Bath Process of Whitening and Dyeing of Fluorescent Acrylic Yarns", Journal of Textile Research, Key Laboratory of Advanced Textile Materials and Manufacturing Technology, Ministry of Education, Zhejiang Sci-Tech University, Hangzhou, 2008.

11. H. Panda, "Proposed Technology of Textiles Dyes & Pigments," National Institute of Industrial Research, Delhi, India, pp. 247-253.

12. E. R. Trotman, "Dyeing and Chemical Technology of Textile Fibers," 4th Edition, Charles Griffin & Company Ltd, London, 1970, pp. 190-197.

13. ASTM D 1776, "Standard Practice for Conditioning Textiles for Testing," American Society for Testing and Materials, Annual Book of ASTM Standards, Vol. 7, No. 1, ASTM International, West Conshohocken, PA, USA, 2008.

14. AATCC Test Method 61, "Colorfastness to Laundering, Home and Commercial: Accelerated," American Association of Tex Research Triangle Park, N.C., USA, 2010.

Chapter 5

Bacterial Influence on Textile Wastewater Decolorization

Aamr Alalewi, Cuiling Jiang

[1]Department of Hydrology and Water Resources, Hohai University, Nanjing, China; [2]State Key Laboratory of Hydrology-Water Resources and Hydraulic Engineering, Hohai University, Nanjing, China.

ABSTRACT

The study aims to isolate and optimize bacterial strains having the ability to degrade and decolorize azo dyes produced in the final effluent of textile dying industries. In this regard, ten bacterial strains were isolated from wastewater treatment plants, and most of them were subjected to the colored effluents resulting from dilapidated houses. The ability of these bacterial isolations to use a wide range of azo dyes to determine the sole carbon source was determined. According to these screening testes, two bacterial isolations were selected as the most potent decolorizer for azo dyes, and they were identified as Comamanas acidovorns-TN1 and Burkholdera cepace-TN5. The optimization process started with the addition of 1 g/l yeast extract, where the decolorization ability of the two strains increased sharply and according to this experiment, the two azo dyes, Acid orange 7 and Direct blue 75, were selected to complete the study. The effect of different conditional and chemical factors on the decolorization process of Acid orange 7 and Direct blue 75 by Comamanas acidovorns-TN1 and Burkholdera cepace-TN5 was studied. Factors that contributed to the difference were different pH, temperature, incubation period, inoculum size, carbon source, nitrogen source and the respective concentrations of yeast extract. This study recommends the application of the two most potent bacterial strains in the decolorization of the azo dyes, along with acid orange 7 and direct blue 75, specifically in the industrial effluents under all nutritional and environmental conditions.

Keywords: Decolorization; Biodegradation; Azo dys; textile wastewater

1. INTRODUCTION

Environmental pollution has been identified as a major problem in the modern world. The increasing demand for drinkable water, and its dwindling supply, has made the treatment and reuse of industrial effluents an attractive option. One of the most important environmental pollution problems is the color in water courses; although some of these colors are normally present and of "natural" origins (e.g. the color originates from the activity of microorganisms in ponds), a considerable proportion, especially in the lower reaches of rivers that drain large industrial conurbations, originates from industrial effluents. Some colored effluents are associated with the production and use of dyes.

Azo dyes, the largest chemical class of dyes with the greatest variety of colors, have been used extensively for textile, dyeing, and paper painting [1]. These dyes cannot be easily degraded, and some are toxic to higher animals. Over 7×10^5 metric tones of synthetic dyes are produced worldwide every year for dyeing and printing, and out of this, about 5% - 10% are discharged with wastewater. The amount of dye lost depends on the class of dye applied; it varies from 2% loss with the use of basic dyes to about 50% loss in certain reactive sulfonated dyes [2]. The presence of dyes in aqueous ecosystem diminishes photosynthesis by impeding light penetration into deeper layers thereby deteriorating water quality and lowering the gas solubility. Furthermore, the dyes and/or their degraded by-products may be toxic to flora and fauna [3]. Azo dyes consist of a diazotized amine coupled with an amine or phenol, and contain one or more azo linkages. At least 300 different varieties of azo dyes are extensively used in the textile, paper, food, cosmetics and pharmaceutical industries. The effects of pH, temperature, type and concentration of respiration substrates, and oxygen tension on the rate of biological reduction of a variety of azo dyes have previously been investigated [4, 5]. Several combinations of treatment methods have been developed so far in order to effectively process cottontextile wastewater, with decolorization among the main goals for these processes. Chemical coagulation/flocculation techniques, usually combined with activated sludge treatment, have been among the most common processing methods mainly due to the ease of their application. However, the above methods require high amounts of raw materials (coagulants) and also yield large amount of waste solids, leading to the elevation of the total treatment cost. Advanced oxidation processes (Ozonation, UV/H_2O_2) are based on the generation of hydroxyl radicals, which are mainly highly reactive oxidants. They are environmental friendly techniques, since no solid wastes are produced. However, they are not cost effective due to the high consumption of both energy and raw material (e.g.

hydrogen peroxide). Active carbon adsorption and nanofiltration techniques are able to remove dyes from wastewater. Although, the main disadvantage of these methods is the production of a secondary waste stream (or waste solid) that requires further treatment or disposal.

The anaerobic reduction of azo dyes to simpler compounds has been well researched [6-8]. These and other studies have all demonstrated the ability of anaerobic microbes and sludge to effectively reduce azo dyes to their intermediate structures, thus destroying the apparent color. Many of these intermediates are aromatic amines with constituent side groups. By reducing the dye compounds to their intermediates, the problem of aesthetic pollution is accounted for, but a larger and more deleterious problem may be created. Most azo dyes are nontoxic, but a high percentage of their intermediates have been identified as carcinogens. Because of the toxic potential of many aromatic amines, further degradation of the dye compound is necessary if toxicity is to be eliminated or reduced [9,10]. In light of the presumptive evidence, this investigation was designated to fulfill the following objectives:

1) investigate the potential of bacterial isolations from textile industries wastewater and drains (textile effluent adapted bacteria) in some azo dye biodegradation;

2) analyze the bioremediation of secondary metabolites of azo dye degradation (secondary aromatic amines) to environmental friendly non toxic form;

3) how to use cost-effective raw materials (as agricultural and industrial wastes) in the dye biodegradation process.

2. METHODOLOGY

2.1. Isolation and Cultivation of the Most Efficient Decolorizing Bacteria

Six active sludge samples were obtained from three wastewater treatment plants. All samples were transported to the laboratory and screened to obtain the dye decolorizing organisms. All bacterial isolations were cultivated on mineral salts basic medium with the following composition (g/l): Na_2HPO_4, 2.13; KH_2PO_4, 1.3; NH_4Cl, 0.5; $MgSO_4$, 0.2; 1 L tap water and 1ml trace element solution per liter. The trace element solution had the following composition (g/l): $MgSO_4 \cdot 7H_2O$, 7.12; $ZnSO_4 \cdot 7H_2O$0.044; $MnSO_4 \cdot 4H_2O$, 0.081; $CuSO_4 \cdot 5H_2O$, 0.0782; $Na_2MoO_4 \cdot 2H_2O$, 0.025; $FeSO_4 \cdot 7H_2O$, 0.498; Boric acid 0.1 + 0.27 ml of H_2SO_4. The final PH was adjusted at 7.0. The mineral salts medium was supplemented with 1 (g/l) yeast extract. The activated sludge samples (2.5 ml) were transferred into 250 ml flasks containing 50 ml mineral salts medium and dye (prepared using a mixture of 9 types of dyes). All dyes ware mixed together to get stock solution of mixture of dyes 0.9 (g/l) (0.1 g/l of each dye). The stock solution was supplemented with a basic salt medium to get a final concentration of 0.1 (g/l). Each flask contained 50 ml of a sterile mineral salts liquid medium mixed with azo dyes mixture dispensed with one gram of yeast extract. Each flask was inoculated with 2.5 ml of activated sludge. All were incubated in a rotary incubator at 150 rpm for 24 hours at 30°C. Similar flasks were prepared and inoculated with 1 ml of the content of the flasks of the first group. These were incubated for 24 hours. This step was repeated three times in a period of 72 hours. The last (third) group of flasks was incubated for five days. These flasks were used to isolate the target microorganisms by streaking a mineral salts agar medium containing the same ingredients of the previous broth medium, plus agar and 100 ppm of an azo dyes mixture. Separate colonies of the predominant types of microorganisms were purified by re-streaking on the same medium. The purified isolates were examined microscopically to check for purity. Obtained pure cultures were maintained on nutrient agar at 4°C (in refrigerator) [11-13]. Ten morphologically different isolates were obtained from the previous step and studied for colony morphology.

2.2. Screening Program (Testes) to Select the Most Potent Organisms

First screening was done to test the ability of the purified isolates to utilize different groups of dyes as the sole carbon source. Screening to test the ability of isolated organisms to utilize direct blue 75, direct blue 71, reactive blue 194 and direct red 89 as the sole carbon source was carried out in a mineral salts basic medium used in isolation. Yeast extract was replaced by 0.1 g/l individual dye. Organisms were selected on the basis of their ability to grow and reduce pigmentation under these conditions. Colonies of an over night growth were suspended in normal saline to obtain the optical density of 0.6 at wavelength 610 nm. One milliliter of cell suspension was used to inoculate 100 ml bottles containing 25 ml mineral salts basic medium supplemented with 0.1 g/l individual dye. Bottles were then incubated for seven days at 30°C. A second screening was performed to ensure the ability of selected isolations to utilize different groups of dyes as the sole carbon source.

2.3. Identification of the Most Potent of Azo Dyes-Decolorizing Bacterial Isolates

The two most potent bacterial isolations (TN1 & TN5) having the highest decolorization potentiality were selected to complete the study. They were identified on the basis of cell shape, cell arrangement, relation to oxygen, nutritional characteristics, physiological, and biochemical characteristics in terms of Comamanas acidovorns and Burkholdera cepace.

2.4. Analytical Methods

2.4.1. Decolorization Assay

All samples (2 ml) were centrifuged at 11,000 rpm for 10 min. The supernatant was read at absorbance with maximum (λmax) values using thermo spectronic-Genesys 20 spectrophotometer. The efficiency of color removal was expressed as the percentage of the decolorized dye concentration to that of the initial one, i.e., the difference between the initial dye concentration, Dye(i), and the residual dye concentration, Dye(r), of the sample:

$$\text{Color removal}\ (\%) = \frac{\text{Dye}(\text{i}) - \text{Dye}(\text{r})}{\text{Dye}(\text{i})}$$

2.4.2. Determination of Total Protein

The total protein determination was made according to the method used by Lowry [14] utilizing serum albumin as a standard protein.

2.4.3. Determination of Biochemical Oxygen Demand (BOD)

BOD determination was made according to the standard methods for examination of water and waste water (20th edition).

2.4.4. Determination of Chemical Oxygen Demand (COD)

COD determination was made according to a closed reflux, colorimetric method standard method for examination of water and waste water (20th edition).

2.4.5. UV-Visible, Infra Red and HPLC Analysis

Metabolites produced during biodegradation of acid orange 7, direct blue 75, and their mixture were extracted with equal volumes of ethyl acetate. The extract was dried over anhydrous $Na_2SO_4$and evaporated in a rotary evaporator, and then dissolved in ethanol to use for all analyses [15]. The decolorization was monitored using UV-vis spectroscopy analysis (Hitachi U 2800). The active groups of degradation metabolites were monitored by using FT-IR spectroscopy analysis (Jasco, FT-IR 460 PLUS-Japan). HPLC analysis was carried out on a Ceccil model Adept CE 4900 chromatograph equipped with a Cecil model CE 4200 UV detector, an oven column model CE 4601, and a lichrosorb C18 column with a 4.6 mm internal diameter and 25 cm height. A mobile phase composed of 50% methanol, 0.3% H_3PO_4, and 49.7% water was used at a flow rate of 0.5 $ml\cdot min^{-1}$. Elutes were monitored by the UV absorption at 300 nm [16]. In all cases, the parent azo dyes acid orange 7 and direct blue 75 were used as a control.

2.5. Optimization of Decolorization Ability for the Selected Isolates

Addition of Yeast Extract

One g/L yeast extract was supplemented to the mineral salt medium used in screening experiments in a trail to support growth and increase the degradation ability of the selected bacterial isolations. The experiment proceeded in triplicates at pH 7 and an incubation temperature of 30°C in bottles containing 25 ml medium solution.

2.6. Decolorization under Different Culture Conditions

The effect of various culture conditions such as pH, temperature, inoculum size, and incubation period on decolorization of acid orange 7 and direct blue 75 by Com. acidovorns-TN1 and Bur. Cepacia-TN5 was examined.

2.6.1. Effect of PH on the Decolorization of Azo Dyes Acid Orange 7 and Direct Blue 75 by Com. Acidovorns-TN1 and Bur. Cepace-TN5

Colonies of an overnight growth were suspended in normal saline solution to obtain optical density of 0.6 g/l at 610 nm wavelength. One milliliter of cell suspension was used to inoculate 100 ml bottles containing a 25 ml mineral salt basic medium, supplemented with 0.1 g/l individual dye, and 1 g/l yeast extract. The medium was adjusted to pH of 4, 6, 7, 8 and 9 using (1 N) hydrochloric acid (HCl) and (1 N) sodium hydroxide (NaOH). Bottles were incubated for seven days at 30°C.

2.6.2. Effect of Different Incubation Temperature on the Decolorization of Azo Dyes Acid Orange 7 and Direct Blue 75 by Com. Acidovorns TN1 and Bur. Cepace TN5

The experiment was carried out in 100 ml bottles containing 25 ml mineral salt basic medium supplemented with 0.1 g/l individual dye and 1 g/l yeast extract. The medium was adjusted to pH 8 and each bottle was inoculated with a predetermined equal cell density for the two strains. Bottles were divided to be incubated at different temperature: 10°C, 25°C, 30°C, 35°C and 40°C.

2.6.3. Effect of Different Incubation Periods on the Decolorization of Azo Dyes Acid Orange 7 and Direct Blue 75 by Com. Acidovorns-TN1 and Bur. Cepace-TN5

This experiment was carried out in order to investigate the effect of different incubation periods on the decolorization process of the two azo dyes by Com. acidovorns-TN1 and Bur. cepace-TN5. The two strains were allowed to grow in the two azo dyes under the optimum conditions determined from the previous experiments, and were then incubated for 6, 12, 24, 48, 72, 120 and 168 hours, respectively. At the end of each incubation period, azo dyes decolorization (%) and the protein content were assayed.

2.6.4. Effect of Different Inoculum Sizes on the Decolorization of Azo Dyes, Acid Orange 7 and Direct Blue 75, by Com. Acidovorns-TN1 and Bur. Cepace-TN5

Different inoculation sizes of heavy cell suspension of the two bacterial isolates Com. acidovorns-TN1 and Bur. cepace-TN5 (prepared by washing each slant with 20 ml of sterile saline solution under aseptic conditions and optical density was adjusted to obtain an optical density of 0.6 at 610 nm wavelength) were used. The following inocula sizes were applied via 0.2, 0.5, 1, 2 and 3 ml per flask. All other optimal culture conditions were taken into consideration. At the end of incubation period the azo dyes biodegradation were determined for each flask as previously mentioned.

2.6.5. Effect of Different Carbon Sources on the Decolorization of Azo Dyes Acid Orange 7 and Direct Blue 75 by Com. Acidovorns-TN1 and Bur. Cepace-TN5

Different carbon sources were introduced into the two azo dyes mineral salts media at an equimolecular level located at 0.5 g/l. A parallel experiment was made without sugar as a control. The carbon sources were represented by sucrose, glucose, maltose, starch, dextrin, fructose, manitole and lactose. In all cases, other previously mentioned optimal conditions were taken into consideration.

2.6.6. Effect of Different Nitrogen Sources on the Decolorization of Azo Dyes Acid Orange 7 and Direct Blue 75 by Com. Acidovorns-TN1 and Bur. Cepace-TN5

Biodegradation of mineral salts media were supplemented with different nitrogen sources with equivalent amounts of nitrogen present in 0.5 g/l NH_4Cl. The particular nitrogen sources were KNO_3, $NaNO_3$, NH_4Cl, NH_4NO_3, $(NH_4)_2NO_3$, Amm.SO_4, Amm.oxalate, Amm. Molybdenate, and peptone. All other optimal factors were carried out as previously mentioned. The azo dyes biodegradation were assayed for each separate nitrogen source as previously mentioned.

2.6.7. Effect of different Yeast Extract Concentrations on the Decolorization of Azo Dyes Acid Orange 7 and Direct Blue 75 by Com. acidovorns-TN1 and Bur. Cepace-TN5

This test was performed to investigate the effect of different concentrations of yeast extract on the azo dyes biodegradation by Com. acidovorns-TN1 and Bur. cepaceTN5. This experiment was carried out by applying different concentrations of yeast extract (0.1, 0.25, 0.5, 1 and 2 g/l). All concentrations were proceeded as previously mentioned optimal conditions.

2.6.8. Effect of Different Incubation Types (Oxygen Concentrations) on the Decolorization of Azo Dyes Acid Orange 7 and Direct Blue 75 by Com. Acidovorns-TN1 and Bur. Cepacia-TN5

This experiment was carried out in order to investigate the effect of incubation conditions on the biodegradation of azo dyes acid orange 7 and direct blue 75 by Com. Acidovorns-TN1 and Bur. cepace-TN5. This was carried out by incubating the flasks containing the mineral salts media with each azo dye in addition to the optimal sources of carbon and nitrogen resulting from the previous optimization in four ways: static incubation, shaking incubation at 150 rpm, anaerobic and aerobic-anaerobic incubation. At the end of incubation period, azo dyes decolorization were assayed as previously mentioned.

2.6.9. Effect of Consortium as Com. Acidovorns-TN1 + Bur. Cepace.-TN5 on the Decolorization of Azo Dyes Acid Orange 7 and Direct Blue 75

This experiment was carried out in order to investigate the effect of a consortium of Com. acidovorns-TN1 + Bur. cepace-TN5 on enhancing biodegradation of azo dyes acid orange 7 and direct blue 75. This was carried out by inoculation of the two strains, Com. acidovorns-TN1 and Bur. cepace-TN5, into biodegradation media and incubation under all the optimal nutritional and environmental conditions. At the end of incubation period, the two azo dyes biodegradation was assayed as previously mentioned.

3. RESULTS

3.1. Isolation, Purification, and Screening Test for Azo Dye Decolorizing Bacteria

Ten bacterial isolations were conducted from an aeration tank and the input effluent of three wastewater treatment plants (Eldwakhlia-Bassiun), as shown in **Table 1**.

Experiments on growth and the dye removal ability of the pure isolations growing in a mineral salts agar medium containing 100 ppm dye revealed that TN1 & TM2 grew rapidly with more visible colonies than others. Ten bacterial isolations were tested for their ability to utilize direct blue 75, direct blue 71, reactive blue 194, and direct red 89 as the sole carbon source in the mineral salts basal medium used for isolation. Data presented graphically in **Figure 1** shows the different capabilities of the ten bacterial isolations pertaining to this study. This effect can clearly be detected by reviewing the protein content values and their relationship to the decolorization percentage concerning different dyes.

Table 1. Code numbers, isolates locality, cell morphology and gram, catalase reaction of the ten bacterial isolates used for azo dye biodegradation process.

No	Code	Isolation locality	Cell morphology	Gram reaction	KOH reaction	Catalase reaction
1	TN1	Aeration tank of Eldwakhliah WWTP	Rods	-	+	+
2	MKD3	Industrial input effluent of Eldwakhliah WWTP	Cocci	-	+	+
3	MKD5	Industrial input effluent of Eldwakhliah WWTP	Rods	-	-	±
4	MKD7	Industrial input effluent of Eldwakhliah WWTP	Rods	-	+	+
5	TN5	Aeration tank of Eldwakhliah WWTP	Rods	-	-	+
6	MM5	Aeration tank of Mahalet marhome WWTP	Rods	-	+	+
7	MM7	Aeration tank of Mahalet marhome WWTP	Rods	-	-	+
8	BO3	Aeration tank of bassuon WWTP	Rods	-	+	+
9	TM7	Aeration tank of Eldwakhliah WWTP	Rods	-	-	+
10	BO4	Aeration tank of bassuon WWTP	Rods	-	+	+

Concerning the second survey, data revealed that direct red 89 and direct blue 75 gave the most promising results among all the four azo dyes under investigation. It is worthwhile to mention that reactive blue 194 failed to fulfill carbon source requirement for all ten bacterial isolations. With regard to the ten purified bacterial isolations under investigation, six were recorded to have given the highest decolorization percentage and protein content. The six bacterial isolates (TN1, MKD7, TN5, BO3, TM7 and BO4) were tested to grow in the same medium containing higher concentrations (500 ppm).

In the second screening, values of the decolorization percentage and protein content were generally decreased, and the best results were attained in the case of TN1& TN5, so these compounds were selected to complete the study **Figure 2**.

In order to test the ability of TN1 & TN5 to deal with wide range of azo dyes, the ability of the two compounds to utilize other members of azo dyes were tested. The new azo dyes were reactive red, mordant brown, acid yellow, acid orange 7 and acid black. Data presented graphically in **Figure 3**showed that there was no great differences between decolorization percentage recorded after one day and that which was recorded after a period of seven days. Reactive red gave the highest degradation percentage and highest protein content by the two compounds.

After confirmatory screening experiments were performed, the mineral salts media was supplemented with 1 g/l yeast extract to enhance the decolorization process.

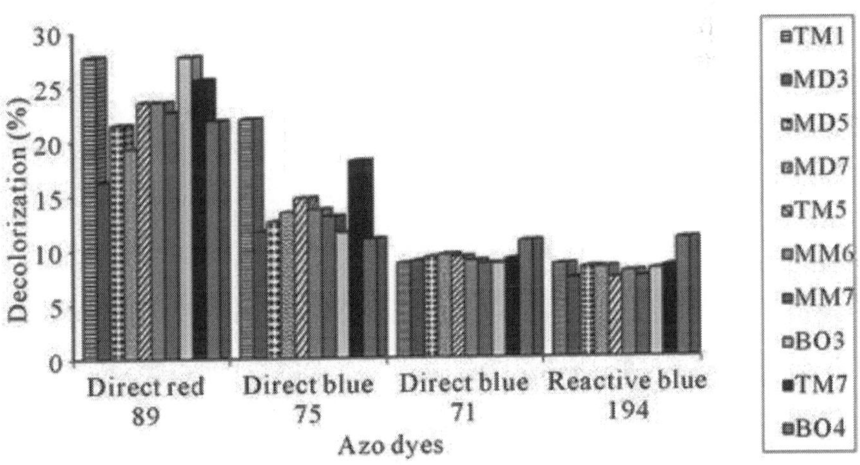

Figure 1. First Survey to test the ability of bacterial isolates to utilutilize azo dyes as sole carbon source.

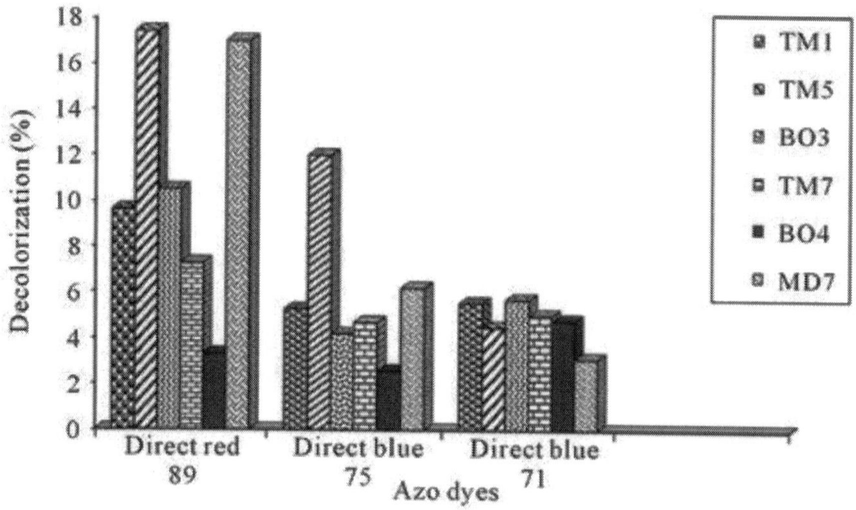

Figure 2. Second confirmatory Survey by the selected six isolates on the same azo dyes.

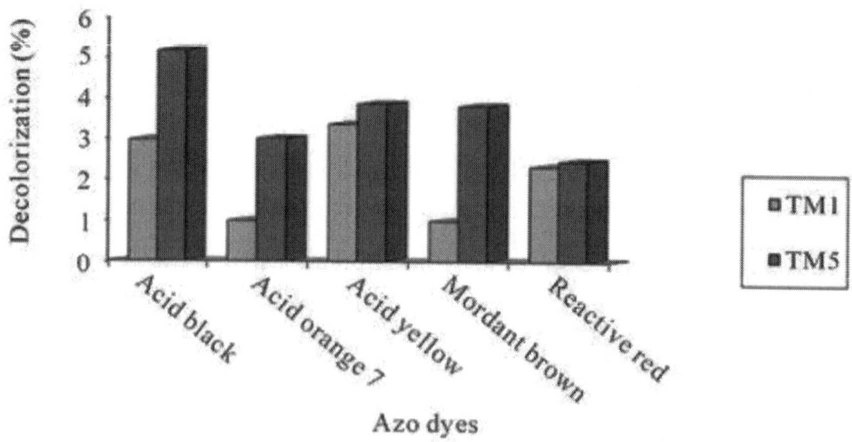

Figure 3. Ability determination of TN1 & TN5 to deal with wide range of azo dyes.

The experiment proceeded to use seven azo dyes: Direct red 89, Direct blue 75, Acid orange 7, Acid yellow, acid black, Reactive red and Mordant brown. The decolorization percentage increased sharply with all dyes but the azo dyes. Direct blue 75 and Acid orange 7 showed the highest results, so they were selected to complete the study. The lowest results for the decolorization process were recorded for Mordant brown and Acid yellow. The lowest protein content was also recorded for compound TN1 in the decolorization of acid yellow and mordant brown. The results are presented graphically in **Figure 4**.

3.2. Parameters Controlling Decolorization Process of Direct Blue 75 & Acid Orange 7 by Com. Acidovorns-TN1 and Bur. Cepace-TN5

Effect of Different Initial pH Values on the Decolorization of Acid Orange 7 and Direct Blue 75 after 24 h and 168 h

pH value affects not only on the decolorization capability, but also on the color stability at pH 9 after autoclaving. Color reduction occurred at pH 9 after autoclaving in the case of both the dyes under study, but there was no effect with this particular pH on the color without autoclaving. The effect of different initial pH values (4 - 9) on decolorization of direct blue 75 & acid orange 7 by Com. acidovorns-TN1 and Bur. cepace-TN5 was recorded after 24 hours and 168 hours, as presented graphically in Figures 5-7 The optimal pH for decolorization by both strains was pH 8. The decolorization percentage was decreased when solution pH made more acidic. This seems to indicate that neutral and slightly basic pH values would be more favorable for the decolorization process of direct blue 75 & acid orange 7 by Com. acidovornsTN1 and Bur. cepace-TN5. Results recorded for decolorization of acid orange 7 by Com. acidovorns-TN1 showed that the decolorization percentage decreased with the decrease in pH after 24 hours and 168 hours, and the optimum pH was 8 for this decrease. However, the optimum pH of growth was 7. On the other hand, results recorded for the decolorization of acid orange 7 by Bur.

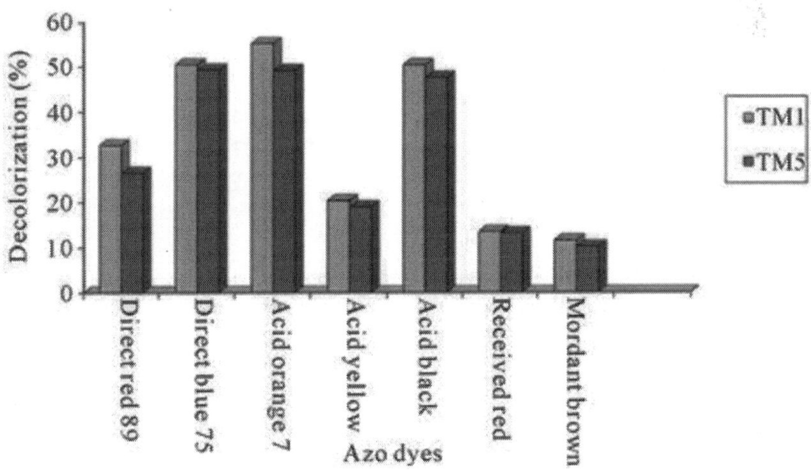

Figure 4. Decolorization of azo dyes by TN1 and TN5 in salts medium supplemented with yeast extract 1 (g/l).

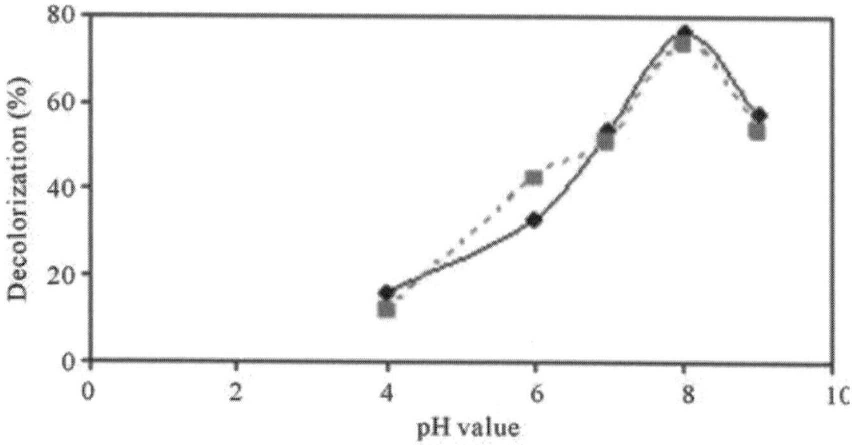

Figure 5. Effect of pH value on decolorization of acid orange 7 by Com. acidovorns-TN1 and Bur. cepace-TN5.

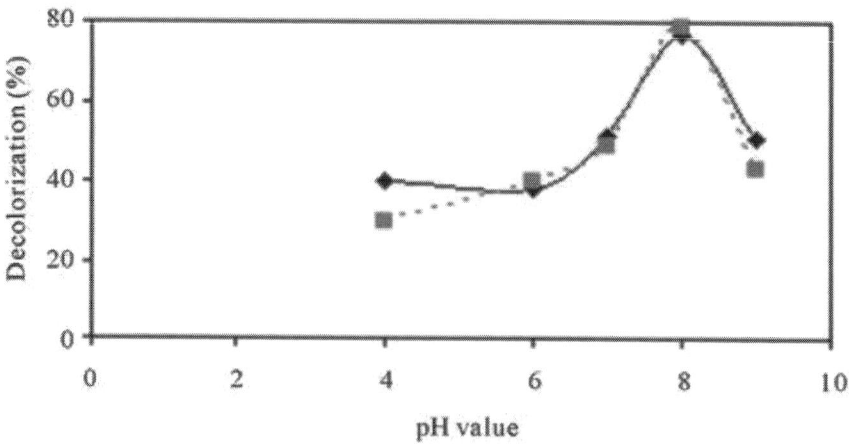

Figure 6. Effect of pH value on decolorization of acid orange 7 by Com. acidovorns-TN1 and Bur. cepace-TN5.

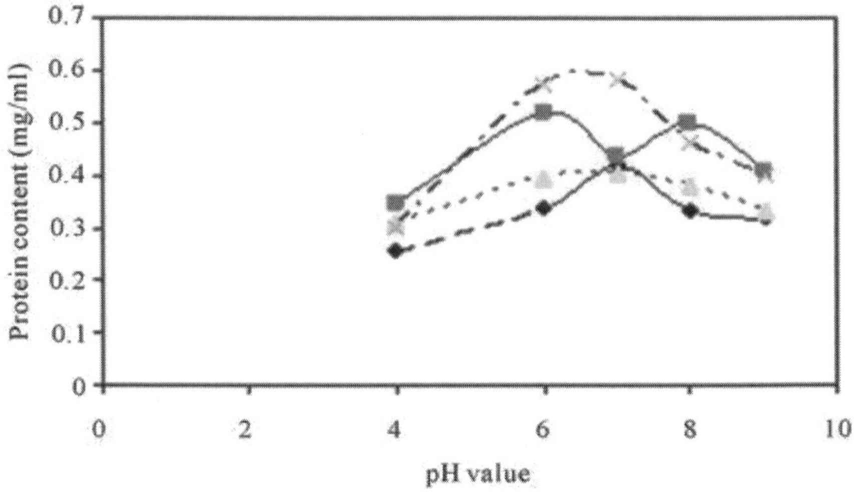

Figure 7. Effect of pH value on protein content of Com. acidovorns-TN1 and Bur. cepace-TN5 in decolorization process of acid orange 7 and direct blue 75.

cepace-TN5 also showed a decrease in the color removal percentage with pH decrease, but this decrease in color removal was not as sharp and optimum growth was at pH 8. The result recorded for decolorization of direct blue 75 by Com. acidovorns-TN1 showed that decolorization percentage decreased with a concurrent decrease in pH after 24 hours and 168 hours. The optimum pH was 8, while the optimum pH of growth was 7 as in the decolorization of acid orange by the same strain. In contrast, decolorization of direct blue 75 by Bur. cepace-TN5 also showed a decrease in the color removal percentage, as the pH decreased, but this decrease in color removal was not nearly as sharp, and optimum growth was still at pH 8 as in the decolorization process in the case of acid orange by the same strain.

3.3. Effect of Different Incubation Temperatures on the Decolorization Process after 24 Hours and 168 Hours

The experiments were performed at different temperatures ranging from 10˚C - 40˚C. The decolorization increased as the temperature increased as evidenced in Figures 8-10. But the color removal ability of both strains decreased sharply at a temperature of 40˚C, and further increases in temperature resulted only in marginal reductions in decolorization activity of the two bacterial strains. The optimal incubation temperature for the decolorization process by the two strains was 35˚C. Decolorization percentage decreased as temperature dropped below 35˚C or increased over this particular value. The growth of the two strains (protein content) also followed the same behavior of decolorization, so the optimum growth temperature of the two strains was determined to be 35˚C. There was no great difference between the decolorization of acid orange 7 by Com. acidovorns-TN1 at temperatures 30˚C and 35˚C. Decolorization of direct blue 75 at temperature 10˚C and 40˚C was nearly the same for Com. acidovorns-TM and Bur. cepace-TN5. The result recorded for decolorization of acid orange by Com. acidovorns-TN1 or Bur. cepace-TN5 showed that the color removal at temperature 10˚C was duplicated at 40˚C after 24 hours and 168 hours.

3.4. Effect of Different Incubation Periods on the Decolorization of Acid Orange 7 & Direct Blue 75 by Com. Acidovorns-TN1 and Bur. Cepace-TN5

Decolorization of AO 7 & direct blue 75 by both bacterial strains was recorded at different intervals of time under optimum pH (pH 8) and optimum temperature (35˚C) conditions, as shown in Figures 11-13. The results recorded for decolorization of acid orange 7 by Com. Acidovorns-TN1 showed that there were slight changes in color removal after six hours and 12 hours of incubation, but at 24 hours the decolorization percentage increased

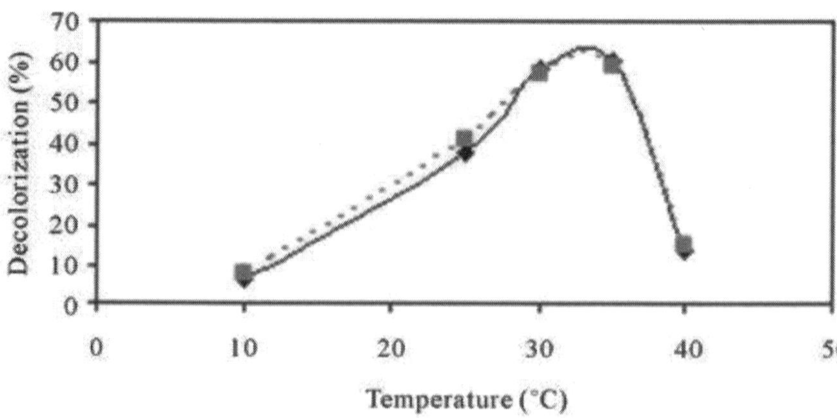

Figure 8. Effect of incubation temperature on decolorization process of acid orange 7 by Com. acidovorns-TN1 and Bur. cepace-TN5.

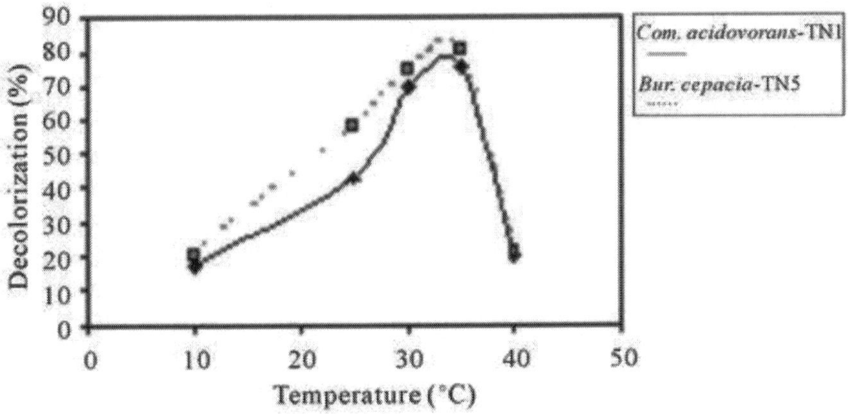

Figure 9. Effect of incubation temperature on decolorization process of direct blue 75 by Com. acidovorns-TN1 and Bur. cepace-TN5.

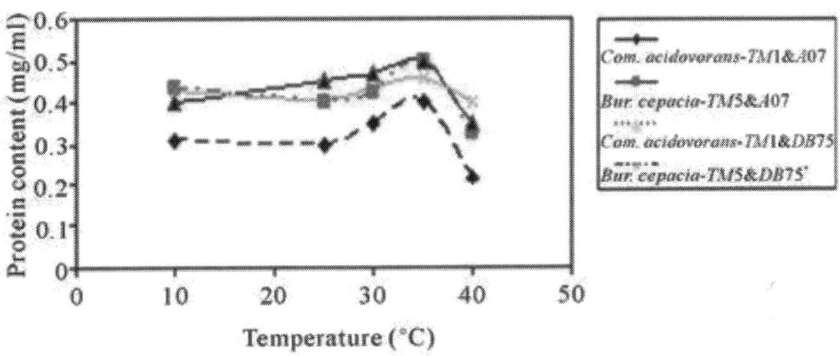

Figure 10. Relation between protein content and incubation temperature in decolorization process of acid orange 7 & direct blue 75 by Com. acidovorns-TN1 and Bur. cepaceTN5.

sharply to reach more than 60% of the total decolorization percentage at the end of incubation. The progress in color removal after 24 hours became slow and the difference between the results recorded at 120 hours and 168 hours was only 1.3% color removal. The growth represented by protein content increased gradually, and not steeply as in decolorization process. On the other hand, the decolorization percentage recorded by Bur. cepace-TN5 with acid orange 7 showed detectable changes in the color removal percentage after 6 hours and 12 hours when compared with the changes in color removal by Com. acidovorns-TN1 during the same interval of time. Protein content measured after 6 and 12 hours for Bur. cepace-TN5 also was higher than that achieved in the same intervals by Com. acidovorns-TN1.

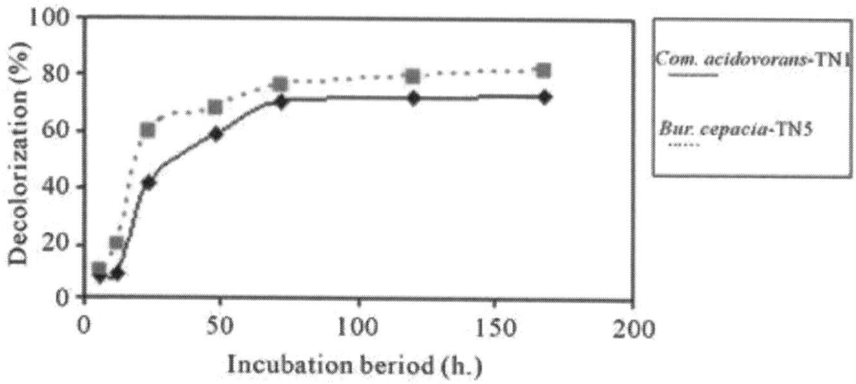

Figure 11. Effect of incubation period on decolorization process ofcid orange 7 by Com. acidovorns-TN1 and Bur. cepace-TN5.

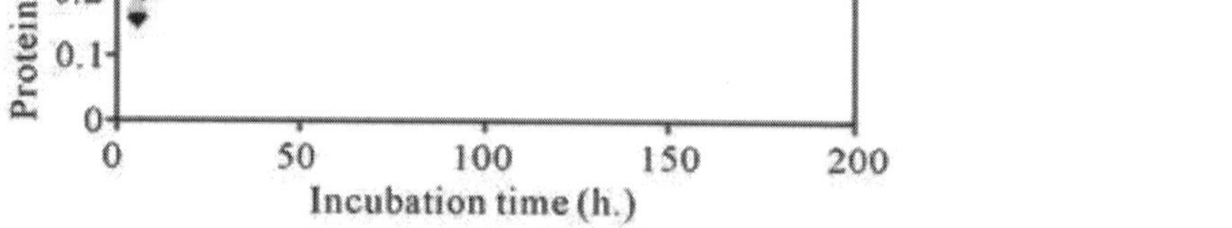

Figure 12. Effect of incubation period on decolorization process of Direct blue 75 by Com. acidovorns-TN1 and Bur. cepace-TN5.

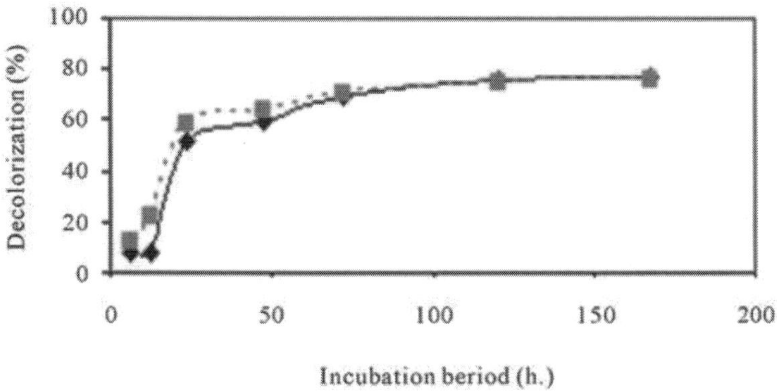

Figure 13. Relation between protein content and incubation period in decolorization process of acid orange 7 & direct blue 75 by Com. acidovorns-TN1 and Bur. cepace-TN5.

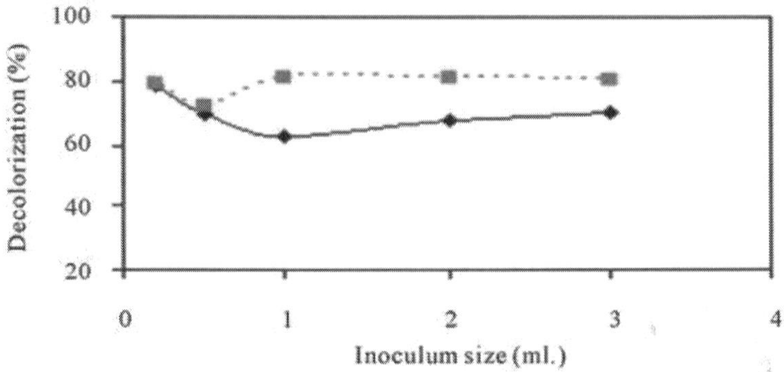

Figure 14. Effect of inoculum size on decolorization process of acid orange 7 by Com. acidovorns-TN1 and Bur. cepaceTN5.

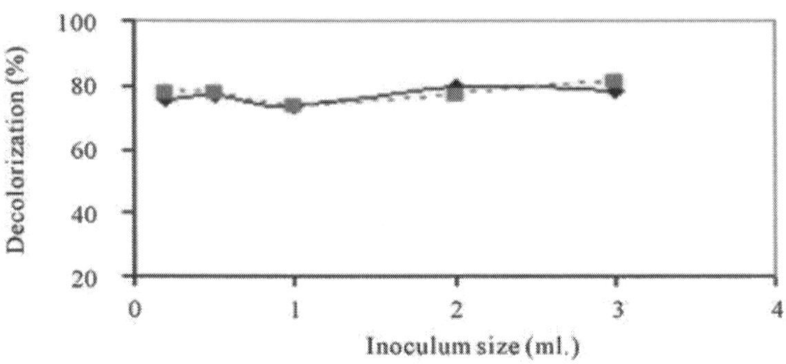

Figure 15. Effect of inoculum size on decolorization process of Direct blue 75 by Com. acidovorns-TN1and Bur. cepaceTN5.

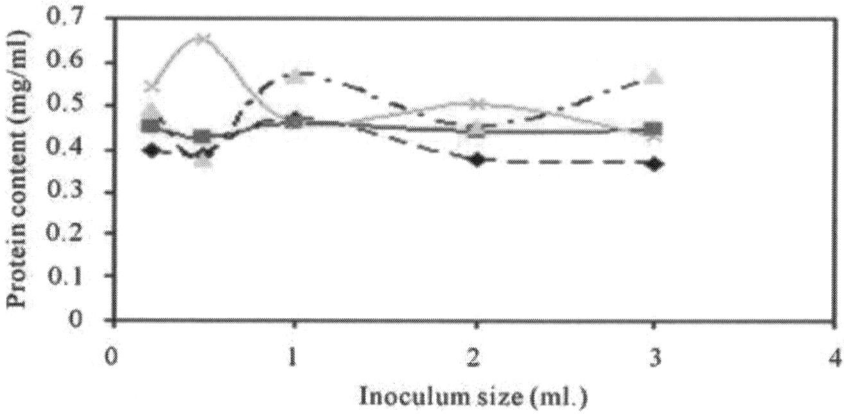

Figure 16. Relation between protein content and inoculum size in decolorization process of acid orange 7 & direct blue 75 by Com. acidovorns-TN1 and Bur. cepace-TN5.

3.5. Effect of Different Inocula Sizes on the Decolorization Process of Acid Orange 7 & Direct Blue 75 by Com. Acidovorns-TN1 and Bur. Cepace-TN5

Results presented graphically in Figures 14-16 indicated that there was no clear relationship detected between the inoculum size and the decolorization percentage after 24 or 120 hours. For instance, the decolorization of acid orange 7 by Com. acidovorns-TN1 after 24 hours did not show detectable differences at inoculum sizes 1 ml, 2 ml, and 3 ml respectively, but decolorization percentage did decrease at inoculum size 0.2 ml and 0.5 ml. The decolorization at 0.2 ml inoculum size was better than that at 0.5 ml, although the color removal achieved after 120 hours using 0.5 ml inoculum size had the highest decolorization percentage recorded in this experiment. The decolorization of acid orange 7 by Bur. cepace-TN5 after 24 hours showed that the best decolorization percentage was at inoculum size (2 ml) and the lowest value was for inoculum size 0.2 ml. Results recorded after 12 hours showed that inoculum size 3 ml gave the highest color removal, and other inocula sizes gave nearly the same value of decolorization except for the inoculum size 1 ml, where the lowest value of decolorization percentage was recorded.

As discussed earlier, there was no clear relationship between the inoculum size and the decolorization percentage after 24 hours or 120 hours. This condition was further proven by the result recorded for the decolorizetion of direct blue 75 by Com. acidovorns-TN1 and Bur. cepace-TN5.

Decolorization of direct blue 75 by Com.acidovoransN1 after 24 hours showed detectable differences at different inocula sizes. Decolorization using inoculum size 0.5 was optimal, followed by decolorization using inoculum size 0.2. The lowest result was recorded for the inoculum size of 3 ml. But, decolorization of direct blue 75 by Com. acidovorns-TN1 after 120 hours indicated that the highest color removal was recorded at the 0.2 ml inoculum size. The relationship between decolorization percentage and inoculum size was absent in the decolorization of direct blue 75 by Bur. cepace-TN5, where it was found that the best decolorization after 24 hours was a inoculum size of 2 ml, but the best result recorded after 120 hours was for the 1ml inoculum size. The recorded protein content for the decolorization of acid orange 7 & direct blue 75 by Com. acidovorns-TM and Bur. Cepace-TN5 had the same behavior, so we are unable to establish a clear relationship between protein content and inoculum size or decolorization percent but in general. The values of protein content of the two strains at any inoculum size with direct blue 75 were higher than those of the two strains at any inoculum size with acid orange 7.

3.6. Effect of Different Carbon Sources on the Decolorization of Acid Orange 7 & Direct Blue 75 by Com. Acidovorns-TN1 and Bur. Cepace-TN5

Textile industrial effluents deficient in carbon content and biodegradation without any extra carbon source is very difficult achieve. Therefore, different co-substrates such as sucrose, glucose, maltose, starch, dextrin, fructose, manitole and lactose (0.5 g/l), were supplemented in the medium and decolorization of all the two dyes (acid orange 7 & direct blue 75) was studied individually by Com. acidovorns-TN1 and Bur. cepace-TN5. Results presented graphically in Figures 17 and 18 revealed that the decolorization of acid orange 7 & direct blue 75 was clearly dependent on the presence of a carbon source. Results recorded for the decolorization of acid orange 7 by Com. acidovorns-TN1 showed that the best decolorization percentage was found to in presence of starch as the carbon source, followed by fructose then maltose. Dextrin and manitole recorded the lowest decolorization percentage. The highest protein content was achieved with glucose and fructose. The lowest protein content was recorded by adding manitole as the carbon source, which is the same as in the decolorization process. Decolorization of acid orange 7 by Bur. cepace-TN5 showed

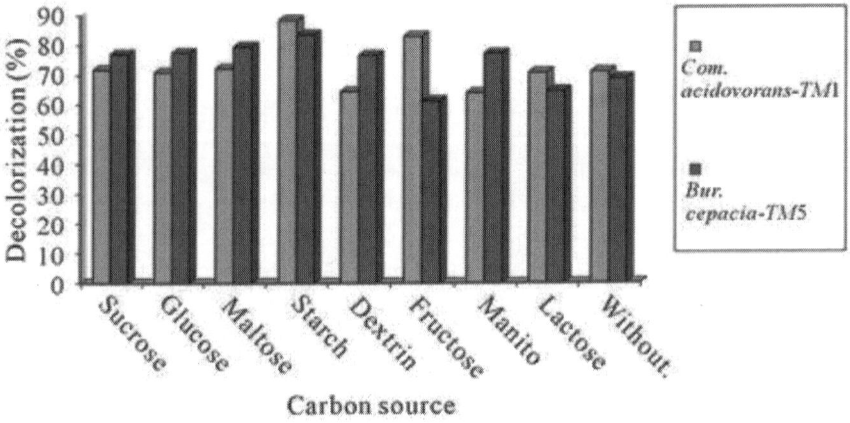

Figure 17. Effect of different carbon sources on decolorization of acid orange by Com. acidovorns-TN1 and Bur. cepace-TN5.

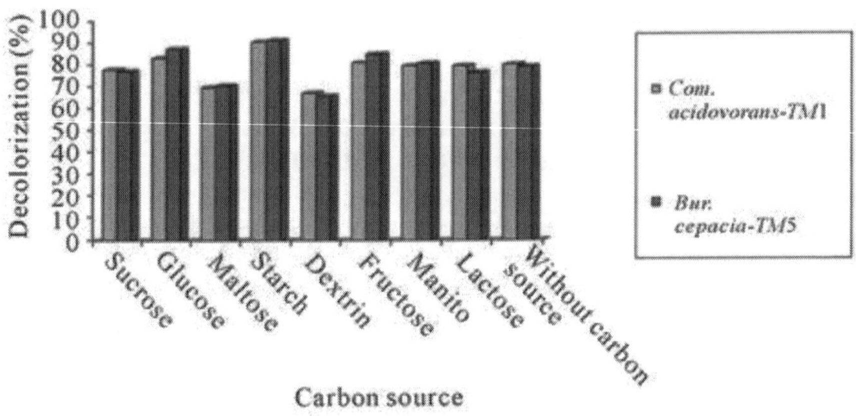

Figure 18. Effect of different carbon source on decolorization of direct blue 75 by Com. acidovorns-TN1 and Bur. cepace-TN5.

that the best decolorization percentage was recorded in the presence of starch as a carbon source, followed by maltose, then glucose. Fructose and lactose recorded the lowest decolorization percentage. The highest protein content was achieved with glucose and fructose as in the decolorization of the same dye by Com. acidovorns-TN1. The lowest protein content was recorded when manitole was used as the carbon source in decolorization process. Starch was also the best co-substrate by using a carbon source in the decolorization of direct blue 75 by Com. acidovorns-TN1 & Bur. cepace-TN5, followed by glucose

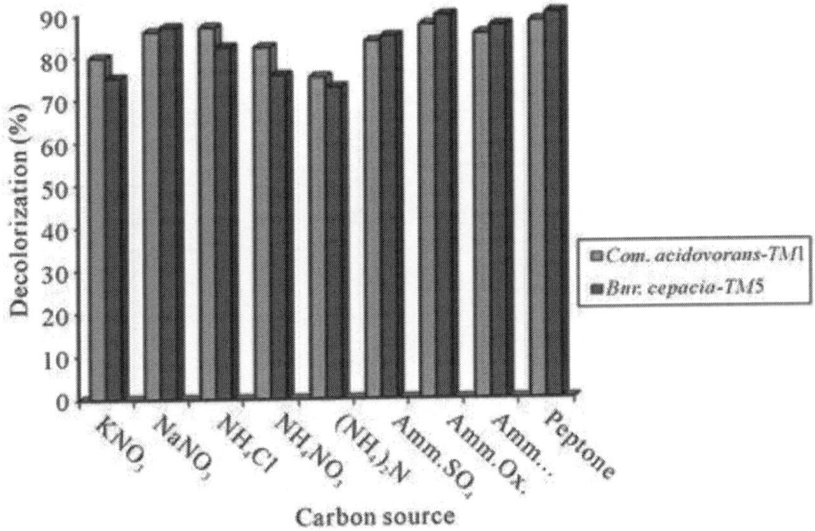

Figure 19. Effect of different nitrogen ources on decolorizetion of acid orange by Com. acidovorns-TN1 and Bur. Cepace-TN5.

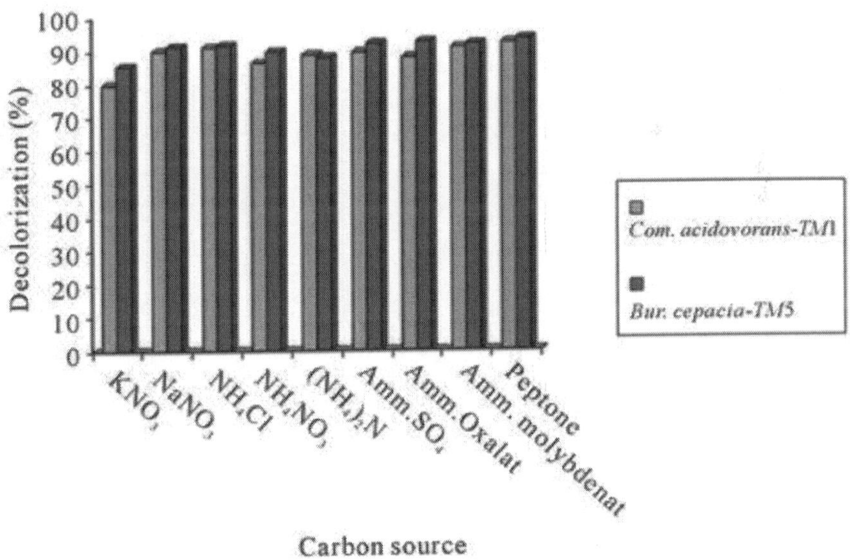

Figure 20. Effect of different nitrogen sources on decolorization of direct blue 75 by Com. acidovorns-TN1 and Bur. Cepace-TN5.

Table 2. A summery for the best conditions resulted from the study for decolorization of acid orange 7 and direct blue 75 by Com. acidovorns-TN1 & Bur. cepace-TN5.

Condition	Com. acidovorns TN1	Bur. cepace TMS
pH	8	8
Temperature	35	35
Incubation period	120 hours	120 hours
Carbon source	Starch	Starch
Nitrogen source	peptone	peptone
Incubation condition	Anaerobic	Anaerobic

and then fructose in the case of decolorization by Com. acidovorns-TN1, and glucose followed by fructose in case of decolorization by Bur. cepace-TN5. The highest protein content was achieved with fructose and dextrin in case decolorization by Com. acidovorns-TN1, and fructose then maltose in case of decolorization by Bur. cepace-TN5. The lowest protein content was by adding sucrose as a carbon source in decolorization of direct blue 75 by each of the two strains respectively.

3.7. Effect of Different Nitrogen Sources on the Decolorization of Acid Orange 7 & Direct Blue 75 by Com. Acidovorns-TN1 and Bur.Cepacia-TN5

A number of organic and inorganic sources of nitrogen were used in this experiment. Results presented graphically in Figures 19 and 20 indicate that, peptone gave the best decolorization percentage and the highest protein content in the decolorization of direct blue 75 or acid orange 7 by each of the two strains respectively. Ammonium oxalate was the next best in the decolorization of each of both dyes by both strains, except in case of decolorization of direct blue 75. Com. acidovorns-TN1 and amm.molebdate were nearly equivalent in color removal ability in the decolorization of direct blue 75 by Com. acidovorns-TN1. The lowest color removal was attained by using $(NH_4)_2HPO_4$ as the nitrogen source in the decolorization of acid orange 7 by each of the two strains individually. The lowest color removal in the case of direct blue 75 was attained by adding potassium nitrate in the decolorization by both of the two strains individually.

We can summarize the optimum condition resulting from the previous experiments as follows in**Table 2**.

4. DISCUSSION

Textile dyeing industries is one of the fastest growing fields and a major export oriented industrial sector, especially in El-Mahalla Elkobra. Azo dye production accounts for more than 1 million tons per year of the world industrial output, and during dying processes, about 40% of this huge amount of azo dyes ends up in drainage. In addition, about 40 - 65 L of textile effluent is generated per kg of cloth produced. On the other hand the release of these compounds into the environment presents serious health problems related to both aesthetic pollution and their toxicity [17]. Colored-dye-wastewater treatment and decolorization presents an arduous task. Wide ranges of pH, salt concentrations and chemical structures often add to the complication. There are many reports on the use of physicochemical methods for color removal from dyes containing effluents [18]. Extensively used coagulation/flocculation techniques produce large amounts of sludge, which requires safe disposal. Partial absorption and membrane filtration techniques lead to secondary waste streams which need further treatment. These constraints have led to the adoption of advanced oxidation processes (AOP) and biological methods as attractive options for the treatment of dye-containing wastewaters. AOP is defined as those processes that use strong oxidizing agents (H_2O_2, Fenton's reagent) or heterogenous photocatalysts, such as TiO_2, ZnO_2, Mn, or Fe in the presence or absence of an irradiation source. These involve mainly the generation of (OH) radicals for the destruction of refractory and hazardous pollutants [19]. These methods do not produce solid waste. However, both AOP and the membrane filtration methods are energy and cost intensive. Among the most economically viable choices available for effluent treatment/decolorization, and also the most practical in terms of manpower requirements and running expenses to adopt and develop, appear to be effect utilization of biological treatments. Fungal, bacterial and yeast decolorization are a promising alternative to replace or supplement present treatment processes. The main objective of the present work is an investigation of the isolation, purification, and screening of bacterial isolations to test whether they have the ability to decolorize and mineralize models of azo dyes, and optimize this ability for application in textile wastewater treatment.

In this regard, ten bacterial strains were isolated from different wastewater treatment plants, most of them exposed to drainage of colored textile effluents in addition to domestic effluents. These bacterial strains were selected after being grown at 30°C and pH 7 in a mineral salt medium supplemented with yeast extract and a mixture of azo dyes at the rotary shaker for 24 hours. This step was repeated in order to permit the bacterial species to have the ability to survive in the presence of the dye.

A screening test for the ability of this isolates to utilize azo dyes as a sole carbon source was established to select the most potent organisms and exclude the chance that decolorization may have occurred due to adsorption alone. This technique was used by [16]. Where the ability of halophilic and halotolerant bacterial strains to utilize remazole blacke B as the sole carbon source was used to select the most effective strains. The screening was done in broth and solid mineral salt media containing different individual dyes to isolate effective organisms in the dyes used to complete this study. Screening tests were conducted three times on different groups of azo dyes to support the selection of organisms which have the ability to utilize different azo dyes as sole carbon sources.

This trial resulted in the fact that only two bacterial isolates TN1 and TN5 were considered to be the best decolorizers of most of the azo dyes used in screening.

The selected two bacterial strains were isolated from the Elmahalla Elkopra wastewater treatment plant, which is exposed to drainage from the textile industry, meaning that TN1 and TN5 were adapted bacterial strains. [20], by studying the textile effluent biodegradation potentialities of textile effluent-adapted and non adapted bacteria, found that there were no great differences in the decolorization ability between adapted and non adapted bacteria. In addition, plasmids were not detected in any of the strains from the effluent adapted or non-adapted sources.

All bacterial strain pellets retained its original color and was not deeply colored because of adsorbed dyes. This indicates that, color removal was due to degradation, and not only to adsorption [21]. The low protein content and low color removal percentage observed in screening tests were attributed to the fact that azo dyes are deficient in carbon content and biodegradation without any extra carbon sources being very difficult. So, optimization experiments were initiated by supplementing the menial salt medium containing dyes with yeast extract. The color removal percentage of most dyes increased sharply after the addition of yeast extract and this is in accordance with other reports [22]. Metabolism of yeast extract is considered essential for the regeneration of NADH, which is the electron donor for the azo bond reduction [16]. The low decolorization percentage of azo dyes, Mordant Brawn, Reactive red and Acid yellow, may due to the fact that decolorization of azo dyes is affected by the permeability of the cell membrane depending upon the molecular weights and intramolecular hydrogen bonds between the azo and hydroxyl groups [23]. Since the two bacterial strains TN1 and TN5 were the most potent decolorizers of the azo dyes under study, they were selected purposely for their identification as well as optimization of their color removal ability of the two azo dyes, Acid orange 7 and Direct blue 75. These colors were also selected to complete the study representing the other azo dyes because they gave the highest color removal ability in screening experiments, and therefore represent the most usable groups of azo dyes. Identification trials indicate that they are related to genus Comamonas and Burkholderia (previously seudomonas), therefore they can be both Com. acidovorns-TN1 and Bur. cepace-TN5. Although there is not sufficient research on using Comamonas sp. and Burkholderia sp. in the decolorization of textile wastewater containing azo dyes, the history of these two species in bioremediation, especially in the degradation of compounds related to unites which azo dyes are constructed (as nitro aromatics), indicate that the use of Comamonas sp and Burkholderia sp in this field is promising. Burkholderia sp. strain AK 5 utilized 4-aminophenol (this compound is an intermediate in the degradation of hydroxyacetanilide and azo dyes) as the sole carbon, nitrogen, and energy sources. A pathway for the metabolism of 4- aminophenol in strain AK-5 was proposed based on the identification of three key metabolites by gas chromatography-mass spectrometry analysis. Strain AK-5 converted 4-aminophenol to 1,2,4-trihydroxybenzene via 1,4-benzenediol. 1,2,4 Trihydroxybenzene 1,2-dioxygenase cleaved the benzene ring of 1,2,4-trihydroxybenzene to form maleylacetic acid. The enzyme showed high dioxygenase activity only for 1,2,4-trihydroxybenzene, with Km and Vmax values of 9.6 _M and 6.8 _mol min_1 mg of protein_1, respectively [24]. Also, Nitroaromatic, a compound widely used as dyes, was depredated by Burkholderia sp. For example, 4-Methyl-5-nitrocatechol (4M5NC) monooxygenase (DntB) from Burkholderia sp. strain DNT catalyzes the second step of 2,4-dinitrotoluene degradation by converting 4M5NC to 2-hydroxy-5-methylquinone, with the concomitant removal of the nitro group.

Pseudomonas is classified as an extremely large, versatile and adaptable class of microorganisms, and it is not surprising that the Pseudomonas species have featured prominently in research involving the degradation of xenobiotic azo compounds. The biodegradation of azo dyes by Pseudomonas cepacia 13NA was investigated by using the dyes, CI Acid Orange 12, CI Acid Orange 20 and CI Acid Red 88. A three-stage continuous culture system (with the flasks arranged in a series) was found to be more successful then a one stage system, with respect to the rate of degradation of the dyes. This was attributed to: 1) An abundance of nutrients in the first stage of the cultivation system which resulted in a rapid increase in the bacterial population, increasing the number of potential azo degrading bacteria; and 2) A consequent paucity of nutrients in the second and third culture flasks which forced the microorganisms to assimilate the azo dyes into their growth, i.e. the azo dyes were degraded under nutrient limited conditions. These results are unusual as a pre-requisite for degradation of azo dyes is usually that a supplemental carbon source is provided to sustain the metabolic activity of the azo reducing microbial population [4].

REFERENCES

1. C. M. Carliell, S. J. Barclay, N. Naidoo, C. A. Buckley, D. A. Mulholland and E. Senior, "Microbial Decolorization of a Reactive Azodye under Anaerobic Conditions," Water SA, Vol. 21, No. 1, 1995, pp. 61-69.

2. N. Dafale, S. Watea, S. Meshram and T. Nandya, "Kinetic Study Approach of Remazol Black-Buse for the Development of Two-Stage Anoxic-Oxic Reactor for Decolorization/Biodegradation of Azo Dyes by Activated Bacterial Consortium," Journal of Hazardous Materials, Vol. 159, No. 2-3, 2008, pp. 319-328.

3. A. M. Talarposhti, T. Donnelly and G. K. Anderson, "Colour Removal from a Simulated Dye Wastewater Using a Two Phase Anaerobic Packed Bed Reactor," Water Research, Vol. 35, No. 2, 2001, pp. 425-432.

4. K. Wuhrmann, K. L. Mechsner and T. H. Kappeler, "Investigation on Rate Determining Factors in the Microbial Reduction of Azo Dyes," Applied Microbiology and Biotechnology, Vol. 9, No. 4, 1980, pp. 325-338.

5. D. Deng, J. Guo, G. Zeng and G. Sun, "Decolorization of Anthraquinone, Triphenylmethane and Azo Dyes by a New Isolated Bacillus Cereus Strain DC11," International Biodeterioration & Biodegradation, Vol. 62, No. 3, 2008, pp. 263-269.

6. D. Brown and P. Laboureur, "The Degradation of Dyestuffs: Part I. Primary Biodegradation under Anaerobic Conditions," Chemosphere, Vol. 12, No. 3, 1983, pp. 397-404.

7. E. Razo-Flores, M. Luijten, B. Donlon, G. Lettinga and J. Field, "Complete Biodegradation of the Azo Dye Azosalicylate under Anaerobic Conditions," Environmental Science and Technology, Vol. 31, No. 7, 1997, pp. 2098- 2103.

8. S. Chinwetkitvanich, M. Tuntoolvest and T. Panswad, "Anaerobic Decolorization of Reactive Dyebath Effluents by a Two Stage UASB System with Tapioca as CoSubstrate," Water Research, Vol. 34, No. 8, 2000, pp. 2223- 2232.

9. W. G. Levine, "Metabolism of Azo Dyes: Implication for Detoxification and Activation," Drug Metabolism Review, Vol. 23, No. 3-4, 1991, pp. 253-309.

10. F. He, W. R. Hu and Y. Z. Li, "Biodegradation Mechanisms and Kinetics of Azo Dye 4BS by a Microbial Consortium," Chemosphere, Vol. 57, No. 4, 2004, pp. 293- 301.

11. N. Hayase, K. Kouno and K. Ushio, "Isolation and Characterization of Aeromonas sp. B-5 Capable of Decolorizing Various Dyes," Journal of Bioscience and Bioengineering, Vol. 90, No. 5, 2000, pp. 570-573.

12. K. Kumar, S. S. Devi, K. Krishnamurthi, S. Gampawar, N. Mishra, G. H. Pandya and T. Chakrabarti, "Decolorization, Biodegradation and Detoxification of Benzidine Based Azo Dyes," Bioresource Technology, Vol. 97, No. 3, 2006, pp. 407-413.

13. B. Chen, "Understanding Decolorization Characteristic of Reactive Azo Dyes by Pseudomonas Luteola: Toxicity and Kinetics," Process Biochemistry, Vol. 38, No. 3, 2002, pp. 437-446.

14. O. H. Lowry, N. J. Rosebrough, L. A. Farr and R. J. Randall, "Protein Determination with the Folin Phenol Reagent," Journal of Biological Chemistry, Vol. 193, 1951, p. 265.

15. D. C. Kalyani, A. A. Telke, R. S. Dhanve and J. P. Jadhav, "Ecofriendly Biodegradation and Detoxification of Reactive Red 2 Textile Dye by Newly Isolated Pseudomonas sp. SUK1," Journal of Hazardous Materials, Vol. 163, No. 2-3, 2009, pp. 735-742.

16. S. Asad, M. A. Amoozegar, A. A. Ppurbabaee, M. N. Sarbolouki and S. M. M. Dastgheib, "Decolorization of Textile Azo Dyes by Newly Isolated Halophilic and Halotolerant Bacteria," Bioresource Technology, Vol. 98, No. 11, 2007, pp. 2082-2088.

17. G. Mezohegyi, A. Kolodkin, U. I. Castro, C. Bengoa, F. Stuber, J. Font and A. Fabregat, "Effective Anaerobic Decolorization of Azo Dye Acid Orange 7 in Continuous Up Flow Packed-Bed Reactor Using Biological Activated Carbon System," Industrial & Engineering Chemistry Research, Vol. 46, No. 21, 2007, pp. 6788-6792.

18. J. H. Churchley, “Removal of Dyewaste Color from Sewage effluent—The Use of a Full Scale Ozone Plant,” Water Science and Technology, Vol. 30, 1994, pp. 275-284.

19. I. A. Alaton, I. A. Balcioglu and D. W. Bahnemann, “Advanced Oxidation of a Reactive Dye Bath Effluent: Comparison of O_3, H_2O_2/UV-C and TiO_2/UV-A Process,” Water Research, Vol. 36, No. 5, 2002, pp. 1143-1154.

20. O. D. Olukanni, A. A. Osuntoki and G. O. Gbenle, “Textile Effluent Biodegradation Potentials of Textile Adapted and Non-Adapted Bacteria,” African Journal of Iotechnology, Vol. 5, No. 20, 2006, pp. 1980-1984.

21. K. C. Chen, W. T. Huang, J. Y. Wu and J. Y. Houng, “Microbial Decolorization of Azo Dyes by Proteus Mirabilis,” Journal of Industrial Microbiology & Biotechnology, Vol. 23, No. 1, 1999, pp. 686-690.

22. K. C. Chen, J. Y. Wu, D. J. Liou and S. J. Hwang, “Decolorization of Textile Dyes by Newly Isolated Bacterial Strains,” Journal of Biotechnology, Vol. 101, No. 1, 2003, pp. 57-68.

23. C. Yatome, T. Ogawa, D. Koda and E. Ikeda, “Biodegradability of Azo and Triphenylmethane Dyes by Pseudomonas Pseudomallei 13NA,” Journal of the Society of Dyers and Colourists, Vol. 97, No. 4, 1981, pp. 166-169.

24. S. Takenaka, S. Okugawa, M. Kadowaki, S. Murakami and K. Aoki1, “The Metabolic Pathway of 4-Aminophenol in Burkholderia sp. Strain AK-5 Differs from That of Aniline and Aniline with C-4 Substituents,” Applied and Environmental Microbiology, Vol. 69, No. 9, 2003, pp. 5410-5413.

Chapter 6

Preparation, Characterization and Application of Ultra-Fine Modified Pigment in Textile Dyeing

Chaoxia Wang[1] and Yunjie Yin[1]

[1] Key Laboratory of Eco-Textile, Ministry of Education, School of Textiles and Clothing, Jiangnan University, Wuxi, China

1. INTRODUCTION

Pigments are kinds of insoluble colorants used for fibers, plastics and other polymeric materials, which can retain stable chemical structure throughout the coloration process in its dispersed solution [1]. The pigments were obtained from mineral materials, vegetable materials, animal waste materials since 1200 BC. The significant development of pigment in textile field from 18th century attributed to a huge expansion in the range of synthetic pigments [2,3]. Currently, pigment perhaps is one of the most commonly and extensively colorations in fiber dyeing, due to its easy application to a variety of fibers and environmental friendly aspects [4].

As a result of the different physical and chemical characteristics between pigments and dyes, dyes are able to dissolve in water and penetrate into the substrate in the soluble form, while pigments are insoluble and difficult to be dispersed in water without the aid of dispersants [1]. The major drawback of pigments is the poor dispersing stability, which is affected by the particle size and surface charge on account of the imperfection of the dispersants [5]. Pigment dispersing system includes two main types according to the medium: solvent-based pigment and water-based pigment.

Solvent-based pigment systems, in which the organic solvent is used as dispersion medium, are more suitable for certain applications [6]. However, it is difficult to obtain pigment particles that are smaller than 100 nm in organic solvents. When the particle size of pigment is large, the interactions between pigments and organic solvents (e.g., hydrogen bonding) can disrupt the interactions with stabilizing dispersants. As a result, distinct particle aggregation that influences optical properties and viscosity occurs. To obtain a stable solvent-based pigment system, appropriate pigment amount, dispersants and solvents are the key factors [7].

Water-based pigment systems are environment-friendly pathways. Pigments are dispersed into water with the aid of auxiliaries, such as dispersants, emulsifiers, anti-setting agents, etc. Water-based pigment has been widely applied in coloration for textiles, paints, architecture, wood and so on [8,9]. But the unmodified pigment dispersion that contains pigments with large particle size is unstable and also suffers from the problems as precipitation and floating color. And then the color, fastness, handle and uniformity of the fabrics dyed with the unmodified pigment dispersion are influenced.

Ultra-fine modified pigment (UMP) is referred to nanometer or micrometer pigment particle by physical and/or chemical modifications in a composite disperse system with attaching functional groups for pigment dispersion. These functional groups often have contribution to appear the vivid color for the chromophoric groups in the pigment particle by selectively reflecting and absorbing certain wavelengths of visible light [6]. The UMP dispersion usually shows relative stability and higher color strength (K/S value) which closely approaches to that of dyes [10]. When UMP is used to color, the finely divided, insoluble particle remains throughout the coloration process [1]. Without the binder or less the amount of binder added in printing paste, the good handle will be achieved. And the better fastness is attained with the stronger attraction between the UMP and the fiber [10,11]. The UMP systems are divided into nonionic pigment dispersion, anionic pigment dispersion and cationic pigment dispersion according to the charged particles. Compared to the nonionic pigment dispersion, the pigment dispersion with ionic dispersant may produce the stronger combining power to the fabric [12,13].

The cationic disperser has been widely used in many fields. It had been reported in Japanese patents that cationic disperser was applied in electrodeposits coat. Being prepared with cationic disperser, the cationic pigment dispersion is able to decrease the migration of the pigment, and favorable dyeing deepness and color fastness are also obtained when such pigment is applied to pretreated cotton fabric and terylene-cotton fabric in pad dyeing [14]. The cationic surfactant is grafted onto the acrylic acid copolymer to attain disperser which is then applied to the anthraquinone pigment to prepare ink for ink-jet printing on paper [15]. In addition, there are many applications of cationic disperser in other fields, such as papermaking, image thermomigration, ink-jet printing paper, inorganic nano-scale powder and preparing hydrophilic ethylene copolymer dispersion. Moreover, pigment dyeing by exhaust process is possible by imparting substantivity with cationic reagents to induce the necessary affinity between pigments and fibers. The chemical modification of fiber with cationic reagents has been carried out to improve the dyeing properties [1,16].

In this chapter, the preparation of UMP, such as dispersing process, grinding process, ultrasonic wave process, microencapsulated process and microfluidics process are all summarized. And the particle size, disperse stability, Zeta potential and color properties, which are used to characterize the UMP are also analyzed. Moreover, the dyeing properties of UMP on cotton, silk, wool and acrylic yarns are mainly reviewed.

2. PREPARATION OF UMP

Different from dyes, UMPs are totally insoluble in organic mediums due to its strong intermolecular aggregation, and they are required to be finely ground and dispersed in an organic medium to warrant their gloss appearance, lighting efficiency and maximizing material utilization. In order to prepare stable UMP, the particle size of UMP needs diminishing sharply to smaller than 1 μm and the process can be achieved via dispersant, microencapsulation, grinding, ultrasonic wave, microfluidizer processes and their combined process [17-19].

The stability of UMP is determined by forces among the UMP particles, when the repulsive force is higher than attractive force, the UMP particles will stably disperse into the media, otherwise, the UMP particles will aggregate together. In UMP with polymer dispersant, the repulsion forces among particles including static-electronic and steric repulsion are produced by the adsorbed polymer. The former can be measured by the Zeta potentials. The latter is closely connected with the thickness of adsorbed polymer layer [14]. The dispersant and microencapsulation processes are usually utilized together with other methods in the preparation process.

2.1. Dispersing Process

2.1.1. Low molecular weight dispersant

The UMP dispersion is a surface modification technique, typical of which is a chemical bonding of low molecular weight hydrophilic moieties on UMP surfaces. Hydrophilic moieties work by electrostatic interaction between UMP particles and vehicle solvents. Usually there are no polymeric substances, surfactants or dispersion to aid in surface-modified UMP dispersions, it is necessary to add binder resins before it is applied to textiles [20].

Low molecular weight dispersants are commonly used for wetting and dispersing the UMP particles. The effect of Triton X-100 on the colloidal dispersion stability of CuPc pigment nanoparticles was investigated by Dong et al [18]. The influence of the hydrophobic chain of quaternary dispersers on the properties of pigment dispersion was discussed by Fang et al and good dispersion effects were obtained when the hydrophobic chain were 14 or 16 [21]. The aqueous suspensions of organic pigment particles using cetyltrimethylammonium bromide (CTAB) and sodium dodecylbenzene sulfonate (SDBS) as additives were prepared by Wu et al and a uniform hydrous alumina film could be formed on the organic pigment particle surface with anion surfactant SDBS [22].

But the dispersion system only using low molecular weight dispersants is still suffered from the lack of stability for long-term storage or at high temperature [8], so additional process or dispersant are necessary.

2.1.2. Polymeric dispersant

Polymeric dispersants are a class of specially designed, structured materials and show good properties in the stabilization of organic UMP. In aqueous media, polymeric dispersants are adsorbed onto UMP surface via anchor groups to build voluminous shells or intensify the charges around UMP surface, thereby preventing flocculation and coagulation of the UMP, which can greatly improve the stability of the UMP dispersion [23-25].

The UMP particles are usually quite hydrophobic. In order to achieve a good stabilization in aqueous UMP dispersions, many formulations have been proposed. The application of polymer surfactants in combination with ultrasonic action can significantly improve the quality of dispersed systems. Some aspects concerning UMP-polymer interaction and formation of adsorption layers under mechanical action need additional elucidation. The colloid stabilization of aqueous dispersions with polymer surfactants is believed to be a consequence of adsorption of the amphiphilic macromolecules on the particle surface resulting in mono- or multi-layers of certain structure and thickness which provide certain sterical and/or electrostatic stabilization effects.

Polymer adsorption from aqueous solution on a particle surface is a result of specific interactions of various active sites on the particle surface with corresponding sites (groups) of the macromolecule. Therefore the chemical structures of the stabilizers are believed to be adjusted to the nature of each type of the particles [26]. Fu and his coworkers reported that pigment particles with the diameter of 20-120 nm were uniformly distributing in aqueous media. –COOH of PSMA which encapsulated onto the surface of pigment would build a voluminous shell and also intensify the charges of particles, which could effectively hinder the attraction among particles [27].

2.1.3. Copolymer dispersant

Copolymer dispersants are advantageous for providing multiple anchoring sites toward UMP surface as well as structurally more designable for solvating with the selected solvents. Polymeric structures of random, A-B block, comb-like copolymers prepared by various synthetic techniques have been employed as stabilizers against particle flocculation. However, the methods of anionic and group transfer polymerization are less appropriate since the synthesis of dispersants often involves the monomers with polar functionalities.

Copolymer dispersants are suitable for stabilizing the UMP particles against flocculation during the grinding disruption and storage. The principle for achieving a fine dispersion is a thermodynamically driven interaction among dispersant molecules, UMP particles, and solvents in a collective manner of mutual non-covalent bonding such as electrostatic charge attraction, hydrogen bonding, p-p stacking, dipole-dipole interaction, and van der Waals forces [28].

Copolymer dispersants of high molecular weight have been employed as dispersants to resolve some problems through the molecular designs with multiple anchoring functionalities for interacting with the UMP surface and simultaneously with the involved solvents. The inhomogeneity in geometric shapes of any two nanoparticles may also play an important role for excluding each other from [8]. Recent developments in living/controlled polymerization including nitroxide mediated polymerization (NMP), reversible addition-fragmentation chain transfer (RAFT), and atom transfer radical polymerization (ATRP) have been reported. The copolymers with specific functionalities can be prepared from the monomers with diversified functionalities such as C_1-C_{12} alkyl(meth)acrylate, amine-functionalized (meth)acrylate, and acid-functionalized (meth)acrylate. In addition, copolymer structures can be controlled for their molecular weight distribution and are tailored for specific UMP applications [28].

2.1.4. Siloxane dispersant

Siloxane dispersant can make UMP well dispersed in organic binders due to their hydrophilic/ hydrophobic nature. They convert the hydrophilic surface of UMPs into hydrophobic components which make UMPs compatible with hydrophobic organic resins. Siloxane dispersants can be incorporated very easily into liquids and showed an increase of storage stability, as they tend to deposit more slowly. The dispersing extent and flowability of UMP treated with different siloxane dispersants in water medium are excellent. And siloxane also shows better affinity to UMP than ammonium and nonionic polyether [30]. Siloxane with long alkane chain shows great potential to be a new type of high performance dispersant [29].

Moreover, when the pigment powder modified by siloxane dispersant (10%) is added into the water, the treated sample (b) is able to easily be wetted and enter into the water while the untreated one (a) still floats on the water surface. It is obvious that the siloxane dispersant brings good wettability to Pigment Red CI 170.

2.2. Grinding Process

It is necessary to de-aggregate and de-agglomerate the UMP particles. This is usually accomplished by mechanical action provided by high impact mill equipment, such as the sand mill and ball mill. As the UMP powder is broken down to individual particles by mechanical shear, higher surface areas are exposed to the vehicle and larger amounts of additives are required to wet out newly formed surfaces.

The grinding process can be regarded as a de-flocculation process. In the absence of stabilizing agents, some changes such as reducing K/S value, decreasing gloss and altering rheology probably occur. Grinding process offers UMP in liquids and is suitable for discrete pass as well as for circulation operation. The product passes through a high energy grinding zone inside a grinding chamber and is reduced to the targeted particle size, down to the nanometer range if required. Using a higher specific weight grinding media is likely to reduce the bead size without losing milling energy by use of equal bead filling volumes. The milling process is improved and an optimum grinding result from both a quality and cost perspective can also be achieved by selecting the best bead material, bead size and mill speed from their dependence on the milling product properties [31].

The mechanical grinding of UMPs with the aid of a dispersant is the most convenient method used to produce UMP particles [1]. But mechanical grinding process inevitably has such defects as the relatively large particle size and broad size distribution. The typical size range for particles and/or aggregates produces by traditional mechanical grinding was from 200 nm to 1000 nm. This process often combines with the utilization of ceramic grinding media for particle size reduction. K. Hayashi et al investigated the dry grinding of UMPs with silica particles generates core-shell structures with an average size comparable to the parent silica particles (20 nm) [5].

2.3. Ultrasonic Wave Process

As an environmental and efficient approach, ultrasonic energy was firstly reported to assist various textile processes by Sokolov and Tumansky in 1941 [32]. Sound is transmitted through a medium by inducing vibrational motion of the molecules through which it is travelling. Power ultrasound induces cavitations in liquids, which is the origin of the sound effect in cleaning and in chemical processes [33]. Ultrasound has been widely applied in textiles, for example, preparing nano-scale pigments, fabric pretreatment, dyeing and finishing processes.

Power ultrasound can enhance a wide variety of chemical and physical processes, mainly due to the phenomenon known as cavitation in a liquid medium, which is the growth and explosive collapse of microscopic bubbles. Sudden and explosive collapse of these bubbles can generate hot spots, i.e. local high temperature, high pressure, shock waves, and severe shear force capable of breaking chemical bonds. Dispersion of UMPs involves the application of shear forces to the agglomerates so that they break down into primary particles [34]. If a proper amount of Ultramarine is added to water and stirred briskly, the agglomerates will simply be carried within the flow and hardly change their nature. Ultra-fine modified pigment blue FFG was mixed with dispersant and dispersed with an ultrasonic homogenizer for 30 min, and the particle size was reduced to 91 nm from the original size 220 nm [35].

2.4. Microencapsulated Process

Encapsulating UMP with various polymers is a promising approach for improving the quality of the UMP dispersion. In the last decade, many techniques have been developed for encapsulating UMP [36]. It must be noted that a successful encapsulation technique should not impair the original color appearance of UMP but enhance their dispersion stabilities. Polymeric resins in encapsulation work basically as an adsorbed surface layer around UMP particles and also a fixing agent upon being colored on a substrate. The microencapsulated UMP supplies dispersion stability, and the surface-modified UMP gives fixing on a hydrophilic body, such as the fiber in textiles [20].

The whole encapsulation process probably divides into four steps below [37]. Organic UMP are dispersed into solution of polymeric dispersant leading to the polymeric dispersant absorbing onto the UMP. The absorption auxiliary is added slowly and the copolymers precipitate and encapsulate onto the UMP surface by van der Waals forces. After the precipitate is filtered and dried, the copolymer is tightly encapsulated onto the UMP surface. The copolymer onto the encapsulated UMP is hydrolyzed, and then the UMP is dispersed uniformly in aqueous media.

Dispersion is controlled by attractive force between the UMP and the hydrophobic part of the polymer, and the stability of the particles is made both by electrostatic repulsive force between the particles in water and the polymer-polymer entropic effect. The UMP are dispersed in the solid phase, and then the encapsulation and flocculation are presented. Re-dispersion is necessary to separated microcapsules and finally the complete encapsulation in the medium occurs [20]. The main differences between the surface-modified and the microencapsulated UMP are that the hydrophilic moiety on the surface of particles, i.e. the carboxylic groups in the microencapsulated UMP and the sulfonated groups on the surface of the modified UMP, and additionally the existence of a polymer shell only in the microencapsulated UMP [20,36].

2.5. Microfluidics Process

Microfluidics with excellent dispersing and smashing effects is in favor of leading particles in liquid to reducing particle size to a submicron level to create pure and stable nano-emulsions and suspensions [38]. With this process, the effectiveness of high-performance materials increases, because particles are more uniformly and stably dispersed. Through microfluidizer high shear fluid processors, the fabrics with UMP easily achieve high K/S value and gloss, and the amount of volatile organic compounds by increasing water content is declined. Microfluidizer high shear fluid processors are frequently used to reduce particle size to less than 150 nm.

For example, Pigment Red CI 22 was stirred with an aqueous solution of an anionic polymeric dispersant for 30 min at 10,000 rpm, and the average particle diameter of the obtained UMP dispersion was 1, 500 nm. Whereas, the average particle diameter of 128 nm was available as the dispersion was converted to modified dispersion in a microfluidizer at pressure 22,000 Pa for 2.5 h [13]. Also, a UMP suspension system was prepared by adding the Gemini dispersant and water. Then it was mixed and dissolved and the deformer was added in the mixing process. The Pigment Red CI 22 was added into the above dispersant solution stirring at 600 rpm for 10 min and 9000 rpm for 30 min. The UMP system was treated using a microfluidizer at 172, 368 kPa for 50 min to prepare

UMP suspension system. And particle size of the UMPs was the 211.9 nm, the Zeta potential was 32.4 mV and the viscosity was 1.33 Pa·s [16].

2.6. Combined Process

It is exceedingly difficult to handle a dry powder comprising of UMP particles as the UMP is in the form of larger soft agglomerates. These agglomerates must be broken down into the medium in a process called dispersion. The UMP dispersion can be stabilized by dispersing agents in order to prevent the formation of uncontrolled flocculates. Combining two or more dispersing methods, for example, grinding/ultrasonic wave, microencapsulation/grinding and dispersant/microfluidizer, the homogeneity of making UMP dispersion along with the use of organic dispersants for viscosity control and the prevention of particle from further agglomeration can be realized [8].

Dispersions are usually prepared by ball grinding a mixture of the pigment, dispersant and solvent. This is a simple combined process in which grinding process is used together with dispersant. The ultrasonic process can also disperse the UMP and this method can endow a high-throughput approach. The methacrylic copolymer was used to disperse carbon black in water using both ball mill and ultrasonic approaches. The particle size distribution was determined by laser diffraction. After 16 min both ultrasonification and ball grinding achieve the same particle size distribution (Figure 1). The ultrasonic approach could be used to obtain much smaller sample volumes than the ball mill approach [39].

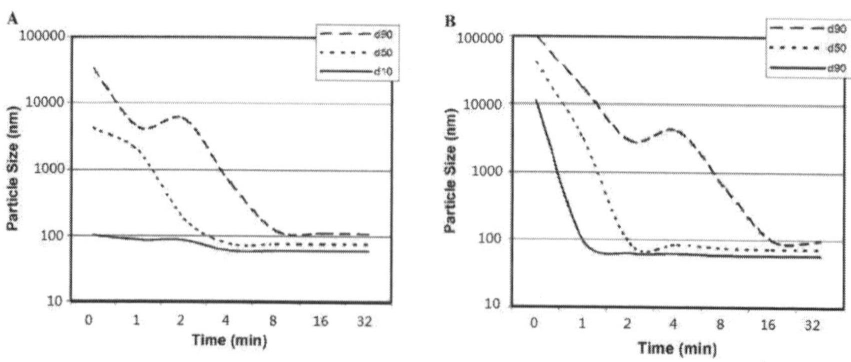

FIGURE 1. (A) Particle size distribution data from ball grinding; (B) Particle size distribution data from ultrasonication

3. CHARACTERIZATION OF UMP

The properties of UMP dispersion can be described by the stability, the adsorbed layer thickness and the Zeta potentials, which are greatly affected by ionic strength, dispersants and dispersing methods. The adsorbed layer thickness and the Zeta potentials can reflect the forces among polymeric dispersants, media, and UMP particles. The characteristics of UMP dispersion affect the applications of the dyed fabric on K/S value, gloss, brightness, and transparency. Some key dispersion characteristics, such as particle size, disperse stability, color properties and Zeta potential properties are investigated as follows.

3.1. Particle Size

UMPs are composed of fine particles which are normally in the submicrometer size range. The UMPs particle size has influence on the color, hide and settling characteristics. Large particles usually settle faster than smaller ones. UMP size and its distribution also influence the light scattering, colloidal stability, appearance and the color properties of the dyed fabric. Therefore, an assessment on the degree of dispersion is necessary to be considered in terms of these critical measurements. In general, color properties, such as strength, transparency, gloss and light fastness of all UMP systems, are affected to a greater or lesser extent by the size and distribution of the UMP particles in the dispersion [1].

UMP particle size and its distribution are greatly influenced by the dosage of dispersants. The particle size decreases at first and then increases with the enhancement of the dispersant, and the particle size reaches its minimum when the ratio is about 10% for phase separation method. The smaller the particle size is, the larger the UMP surface is, thus the more amount of dispersant is needed. However, excessive dispersants, higher than 10%, will dissolve in media instead of attaching onto UMP surface, resulting in increasing viscosity of dispersing media and poor wetting performance in that excess dispersants (Figure 2) [30].

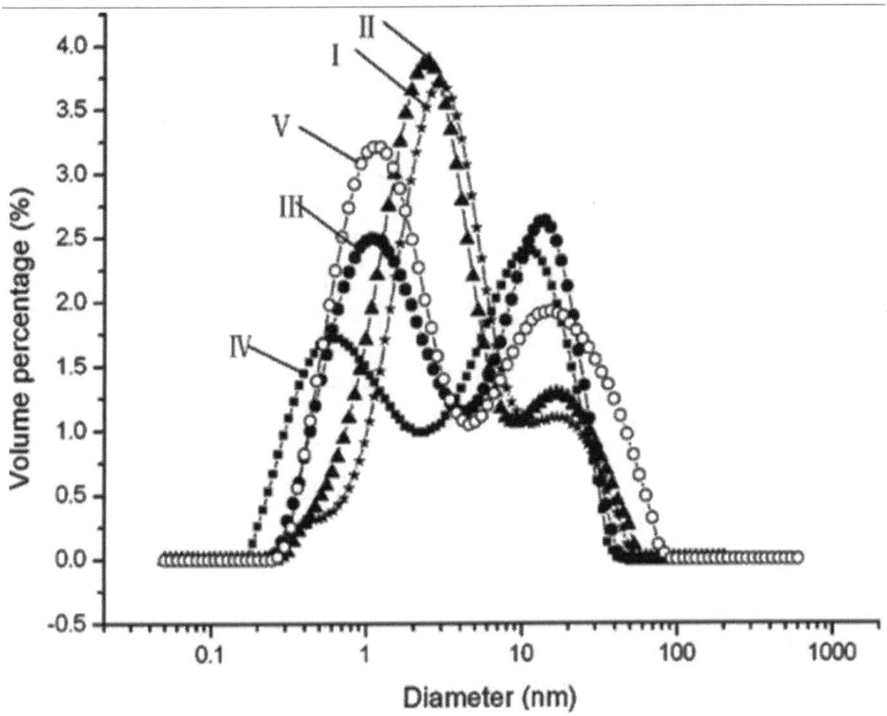

FIGURE 2. Particle size distribution of treated and untreated pigment red 170. (I) 1wt%pigment, 0 wt% dispersant; (II) 1wt% pigment, 0.01wt% dispersant; (III) 1wt% pigment, 0.02wt% dispersant; (IV) 1wt% pigment, 0.1wt% dispersant; (V) 1wt% pigment,10wt% dispersant.

The -COOH group of copolymers encapsulated onto the surface of UMPs would form a polymeric shell around the particles and also increase the surface charges of UMP particles. The interaction between UMP particles is weakened due to the existence of these charges and polymeric layer on the UMP particle surface, so that the UMP particles can be finely dispersed in aqueous media [37]. With the dosage of siloxane dispersant descending, the percentage of big particle ascends at first. However, while the dispersant is greatly excessive (10 wt%), the particle size increases.

Initial experiments, using dilute aqueous dispersions of carbon black of known average-particle size distribution, were performed using both a standard UV/Vis spectrometer and a digital camera to estimate light-transmission/attenuance. The results, summarized in Figure 3(A) and (B), showed a strong relationship between average particle size and light-transmission/attenuance over the size range examined by both these approaches, as demonstrated by the high R^2 values of the polynomial trendline. Because the limited nature of the particle size distributions obtained during these initial experiments, we were only able to demonstrate a correlation over a small size. The digital image method was developed further on account of its potential to allow parallel analysis of multiple samples from the same image [39].

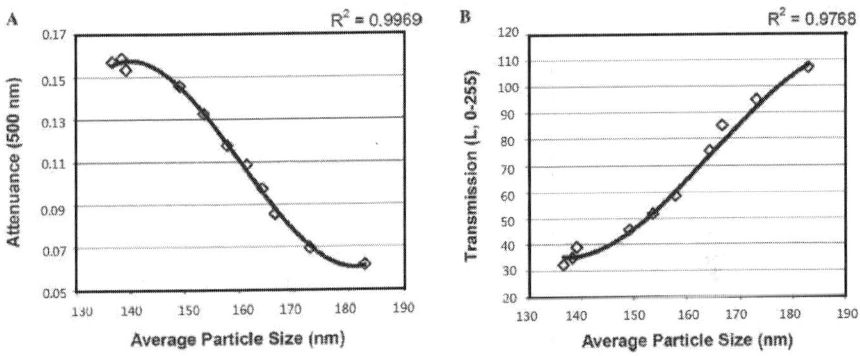

FIGURE 3. (A) Pigment average particle size versus attenuance (D) using UV/Vis analysis. (B) Pigment average particle size versus light-transmission using image analysis

There was a very interesting phenomenon of the particle size distribution of Pigment Red CI 170 with the increase of the cationic siloxane dispersant. In order to clearly describe the distribution process of UMP in water, a experiment was proposed according to the principle of physical chemistry. UMP could not be well dispersed in water with the absence of dispersant. Then, a small amount of dispersant was added into the system. Since the siloxane group of dispersant showed great affinity to the UMP, it tended to anchor on the surface of UMP particle.

3.2. Disperse Stability

Stability of aqueous UMP dispersion is referred to that colloidal properties such as Zeta potentials, viscosity, and storage stability will not change in certain period. It's an important performance of UMP dispersion. Disperse stability can be supported by two generally mechanisms: steric stabilization and charge stabilization. Steric stabilization is due to steric hindrance resulting from the adsorbed dispersing agent, the chains of which become solvated in the liquid medium, thus creating an effective steric barrier that prevents the other particles from approaching too close. Charge stabilization is due to electrical repulsion forces, which are the result of a charged electrical double layer surrounding the particles. The charged electrical double layer developed around the particles extends well into the liquid medium, and since all the particles are surrounded by the same charge (positive or negative), they repel each other when they come into close proximity. The disperse stability usually includes deposited stability, heat stability, pH stability, electrolyte stability and centrifugal stability [35].

Repulsive and attractive forces among UMP particles determine the stability of the UMP dispersion. When the van der Waals forces between UMP particles are higher than steric and static repulsive forces, the UMP particles will combine each other and generate large particles. Instead, the dispersion will stably exist in aqueous media. In aqueous UMP dispersion, polymeric dispersants encapsulated onto the surface of UMPs are ionized in acid solution, and produce some negative charges onto the surface of UMPs to create the static forces between UMP particles. Further, the PSMA encapsulated onto the surface of UMPs can also generate steric repulsion [27]. The whole steric and static forces are larger than van der Waals force between UMP particles, the dispersion will exist in long time, even under centrifugal force or treated at high temperature. It was reported the effect of Triton X-100 on the colloidal dispersion stability of CuPc-U (unsulfonated and hydrophobic copper phthalocyanine) particles and was concluded that the stabilization mechanism for the CuPc-U is inferred to be primarily steric and adding $NaNO_3$ had no obvious effect on the dispersion stability [18].

Stabilization of UMP with polymeric dispersants has been proven to be a good way for the preparation of UMP dispersion with high stability, small particles size, low viscosities, and low moisture sensitivity. In aqueous media, the polymeric dispersant can build polymeric shell around the UMP particles and increase the surface charge on the UMP particles, so that the UMP particles are able to finely dispersed in the aqueous media.

3.3. Zeta Potential

The Zeta potentials of UMP dispersion are analyzed to estimate their disperse stability. The stability of UMP dispersion determined by forces among the UMP particles. When the repulsive force is higher than attractive force, the UMP particles will stably disperse into the media. Otherwise, the UMP particles will arregate together. In the UMP dispersion with polymer dispersant, the repulsion forces among particles produced by the adsorbed polymer include static-electronic and steric repulsion [40]. The former can be measured by the Zeta potentials, and the latter is closely connected with the thickness of

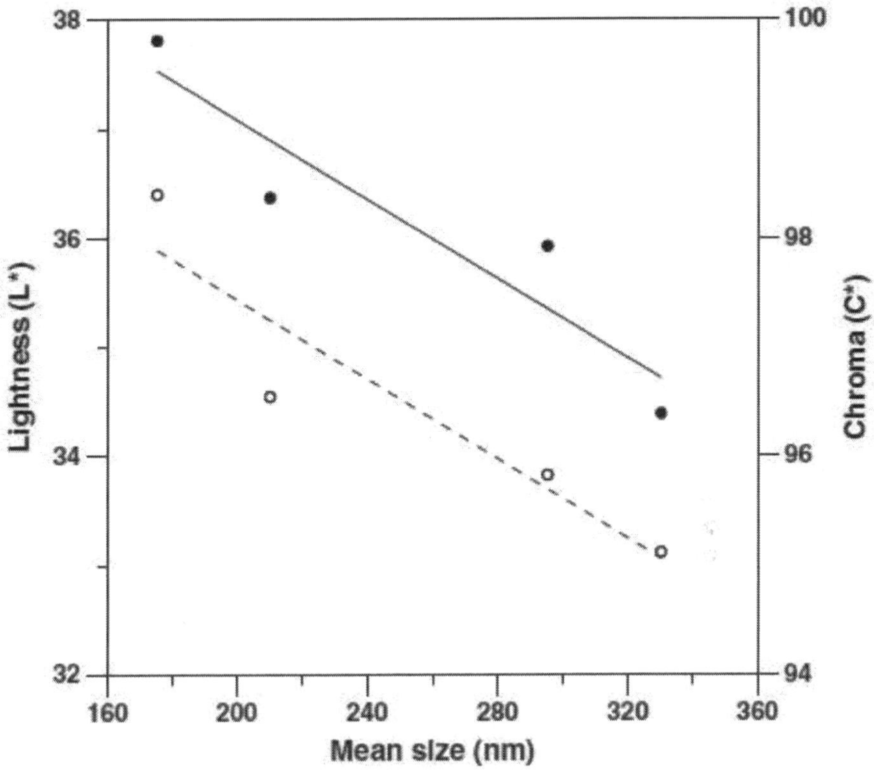

FIGURE 4. The color analyses from the four representative samples of dispersions. (o), Lightness; (•), chroma

adsorbed polymer layer. The studies on the change of Zeta potentials and thickness of adsorbed polymer layer after the addition of solvent are important in understanding the effects of solvent on the stability of UMP dispersion [35,41].

The dispersion mechanism may be interpreted with a pair of electricity layers principle. Although many cationic dispersants gathered on the UMP surface neighbourhood, the quiet balance of positive negative charge didn't change. The cation on the UMP surface adsorbed a great many anions. With cation gathering gradually on the UMP surface, a pair of electricity layers formed. The particles stably existed in dispersion system for the electric charge rejected function mutually. An optimum dispersant dosage would exist, which promoted the formation of double electricity layers [16,42].

3.4. Color Properties

UMP particles which are dispersed finely in dispersions enhance the color depth of UMP dispersions (e.g. transmittance, chroma, and lightness) and improve color display performance. The transmittances were enhanced with decreasing the mean size of dispersion (Figure 4). A color analysis program estimated the lightness (L*) and chroma (C*) from a spectrum of dispersion samples at a wavelength of 380 to 780 nm based on the L*a*b* Color System and D65 light source. The finer sizes of UMP particles in dispersion resulted in the higher lightness (L*) and chroma (C*). Serious aggregation of UMP particles without a dispersant in the dispersion would be performed. Holding the supercritical fluid-assisted dispersion process at the supercritical region conduced to better dispersion [43].

Color properties are presented the color of the UMP dispersion. The absorbance of UMP hydrous dispersion is slightly larger than that of containing organic solvents. The change is mainly attributed to the disparate polarities of the solutions. In these two solutions, while the polarity of solvent UMP dispersion, which contains more ethanol, is weaker. The color of UMP is aroused by the $\pi \rightarrow \pi^*$ energy transition when the chromophore groups are irradiated. The weak ionization of the UMP increases in a solution with high polarity (such as H_2O), and the electric charge in the conjugated system can transport more easily, which leads to an increases in the absorbance. However, the maximum absorption wavelengths for both systems with or without solvent nearly remain the same. This indicates that the chromophore groups of the UMP are not damaged in the UMP dispersion containing organic solvent and the conjugated systems are essentially not altered, therefore, the color hue of the solvent UMP dispersion remains identical to that of the UMP hydrous dispersion [35].

4. DYEING PROPERTIES

4.1. Dyeing Cotton Fabric With Ump

In order to improve the dyeability with UMP, the cationic pretreatment of fabric is used. It can introduce positively charged sites on cotton fabric. Without pigment modification, there are two inhibiting factors. Firstly, pigments are insoluble and have no affinity for cotton and, secondly, the surfaces of pigment particles are usually negatively charged. UMP dyeing has several noticeable advantages compared to dyes. This method is a short process, simple operation, saving energy consumption and low costs, and matching intuitive easy color imitation, steady color hue, hiding power. Because UMP particles have no affinity for fibers, the UMP dyeing fibers does not exist the selective problem and is appropriate for all spices.

There are many reports on cationic modification of cotton. Karrer researched the cationic pretreatment to the cellulose fiber [44]. Guthrie studied the cationic pretreatment of cellulose fiber and dyed it with acidic dye [45]. The chemical modification of cotton to promote dyeability, light fastness and washing fastness has been researched by Cai [12]. Wang investigated chemical modification of cotton to promote fiber dyeability [46]. Burkinshaw analysized cationic pretreatment of cellulose fiber to improve the dye reactivity [47]. Hauser examined the dyeing behavior of cotton that had been rendered cationic by reaction with 2,3-epoxy propyltrimethylammonium chloride and the result showed that excellent dye yields and color fastness properties were obtained without the use of electrolytes, multiple rinsings or fixation agents [48].

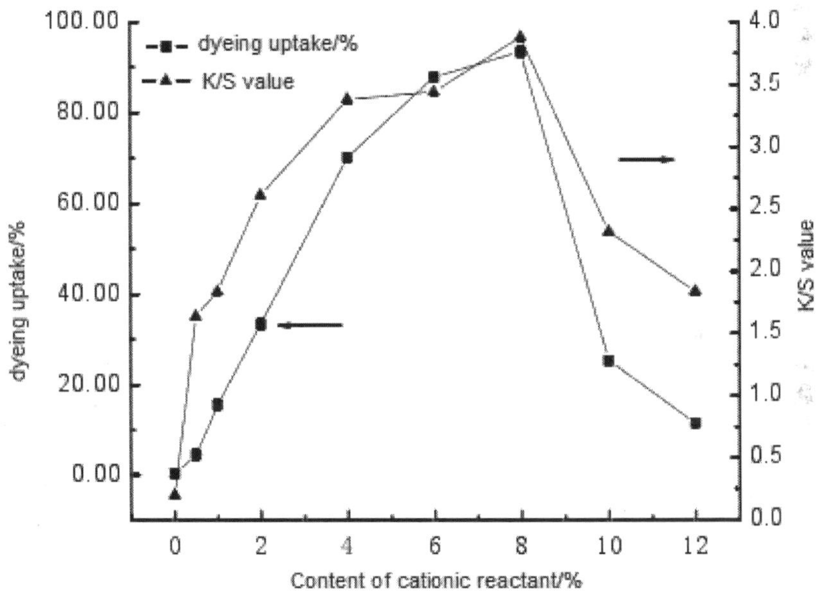

FIGURE 5. Effect of cationic reactant concentration on pigment exhaustion and the K/S value

During exhaustion dyeing process, the dyeing uptake of cotton fabric without modification was very low. Via cationic reagent modification, the uptake of the cationic fabric was increased significantly. Cationic reagent with higher proportions increased the number of positive charges on the cotton fiber, in turn increased pigment uptake.

For example, the cationic polymerization modifier was used to modify the charge of cotton fiber, and this could enhance the uptake of UMP and K/S value with different cationic reagent concentration (Figure 5). The application of cationic polymerization modifier on fibers for cationic modification could reduce the production cost, reduce dyeing temperature, shorten the dyeing process and improve production efficiency [49].

4.2. Dyeing Silk Fabric With Ump

Silk is a natural protein fiber, and the shimmering appearance of silk is due to the triangular prism-like structure of the silk fiber, which allows silk cloth to refract incoming light at different. But the silk fiber also has some defects, such as the problem of bleaching, the loss of pupa protein and the yellowing [50,51].

In order to improve the performance of the silk fiber, silk was modified with cationic dispersant [52]. This improved substantivity of UMP due to the introduction of cationic groups into the silk fabric, and the balance might be due to the adsorption of cationic reagent on surface of silk fabric reaching saturation values. In this situation, more cationic reagent molecules combined with UMP. The bounds between UMP and silk fabric led to a higher K/S values. The different amount of the cationic reagent in pretreatment could impart different amount of positive charge to silk surface, which could affect the K/S values. The UMP was not substantive to silk, the K/S values were low on silk without cationization pretreatment. The K/S values greatly increased and then kept in a balance with the increase of the cationization concentration [53].

Furthermore, penetration property and content of ammonium of silk would also restrict to produce the stronger ionic attraction between the cationic fiber and the anionic UMP. As a result, the adsorption of UMP on silk fabric did not change any more accordingly [54]. Wei reported that after modifying with glycidyltrimethylammonium chloride (EPTAC), the reactions of some amino acids were more rapid than the reaction of glucose. The dyeing ability of modified silk fiber had been improved remarkably. A theory which was called "fixed points" was given to this phenomenon [51]. Wang found that pH in pretreatment affected significantly on the dyeing properties of silk fabric with UMP, because in alkali condition, the chlorohydroxypropyl group of 3-chloro-2-hydroxypropyltrimrthylammonium chloride (CHTAC) converted to an epoxy group and then formed 2, 3-epoxypropyltrimethylammonium chloride which reacted with the nucleophilic amine group in the silk. And The highest K/S values could achieve at pH 8 [53].

With the increase of cationic reagent concentration, the rubbing fastness values appeared increasing trend (Table 1). The washing stain fastness of treated fabric was similarly comparable to the results without fastness improving reagent treatment because the UMP had no affinity with silk without treatment. Compared with untreated silk, the washing change fastnesses of cationic silk increased greatly. This was probably due to the fastness improving reagent acting as the binder between UMP and silk fabric [53].

TABLE 1. Color fastness at different cationic reagent concentration

Fastness improving reagent conc. (g/L)	Rubbing fastness (Grade)		Washing fastness (Grade)		
	Dry	Wet	Color change	Stain cotton	Stain silk
0	2	1-2	2	4	4
10	2-3	2	3-4	4-5	4
20	3	3	4	4-5	4
40	3-4	3	4	4-5	4-5
80	4	3-4	4	4-5	4-5

Cationic pretreatment depended on the extents of reaction between cationic reagent and silk. This might affect the physical properties of the treated fabrics. The tensile strength and elongation at break of fabrics changed with the increase of pH. It was clear that the reaction between cationic reagent and silk took place significantly when pH of the pretreatment solution increased. The tensile strength decreased with the peptide chain hydrolyzation when the pH increased. Also the elongation at break kept slightly decreasing. This revealed that the cationic pretreatment has little effect on the physical properties. The bending rigidity and hysteresis of cationization silk had no change compared with untreated silk. It revealed that the cationization treatment had no impact on the soft properties of fabrics. The handle of fabric decreased a little after dyeing, it might be due to the action of fastness improving reagent.

4.3. Dyeing Wool Fabric With Ump

The wool fabric was modified by cationic reagent and dyed with UMP which was prepared with anionic polymer disperser via exhaust method [55]. The influence of pretreatment conditions such as concentration of cationic reagent, pH value of the bath, temperature and duration of treatment on the dyeing property of the fabric was of importance to the drying properties. With the increase of the cationic content, the K/S value enhanced sharply and then kept a constant value when the content on cationic reagent was higher than 10% (Figure 6). The dry and wet rubbing fastness of the wool fabric dyed with UMP via exhaustion was gray scale ratings of 3-4 and 4, respectively. The tensile strength and elongation at break decreased slightly with the increase of bath pH value. The test results of bending rigidity and hysteresis of bending revealed that the wool fabric had soft handle and good elasticity.

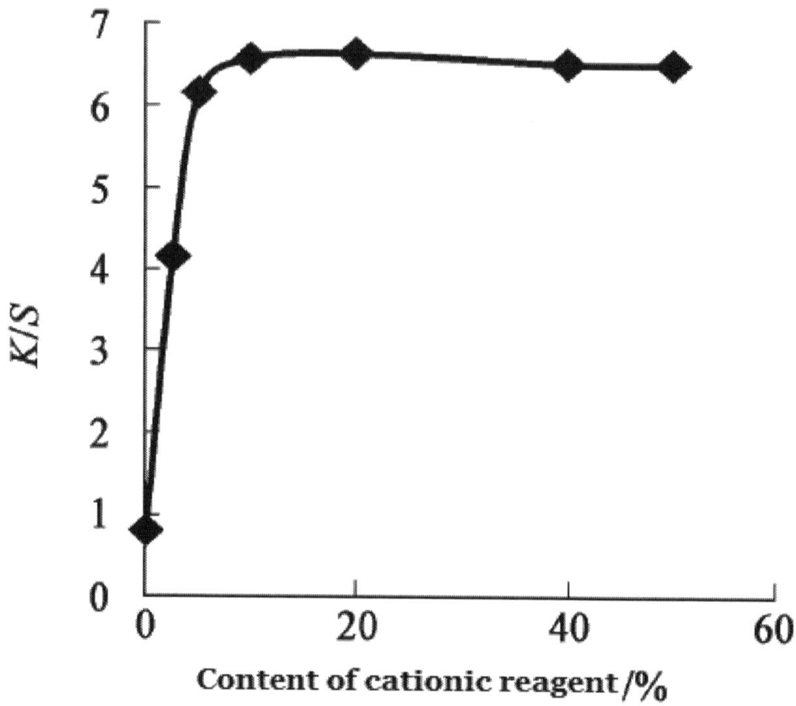

FIGURE 6. The contents of cationic reagent on the K/S value of dyed fabric

4.4. Dyeing Cotton Fabric With Carbon Black Ump

The excellent color and fastness properties make carbon black (CB) particles possible application in textile dyeing. Compared with common sulfur dye, the dyeing process of carbon black UMP can save water and energy due to the shorter dyeing time. However, the dispersion of carbon black UMP in aqueous solutions and adsorption of the particles to cotton fibers are critical in exhaustion dyeing process, because carbon black UMP is hydrophobic and easily aggregate in aqueous solutions.

On the basis of cotton fabrics modified with cationic reagent, the dyeing process and properties of CB dispersions were reviewed (Figure 7). The mechanism of cationic cotton dyed by an exhaustion process using aqueous carbon black UMP dispersions was investigated. Cotton modified with a cationic reagent enhanced the dyeing properties of aqueous carbon black UMP dispersions. There were higher affinity between carbon black UMP and cationic cotton. CB particles could quickly adsorb on the surface of cationic cotton fibers within 5 min and diffuse into the inner pores and cracks on the surface layer of cationic cotton step by step [56]. The cross sections of cationic-modified cotton fibers dyed with different sizes of CB particles were colorless region, and CB particles were only in the cotton fiber grooves, just as "ring-dyeing". After cationic modification, the cotton fiber could have a dark color for the adsorbability of the anionic CB [13].

The color assessment of CB dyed cellulose is correlative to the amount carbon black UMPs on the cotton via electrostatic force. It made possible to compare the charge density difference between nanoparticle ionic materials (NIMs) deposition process and chemical modified process.

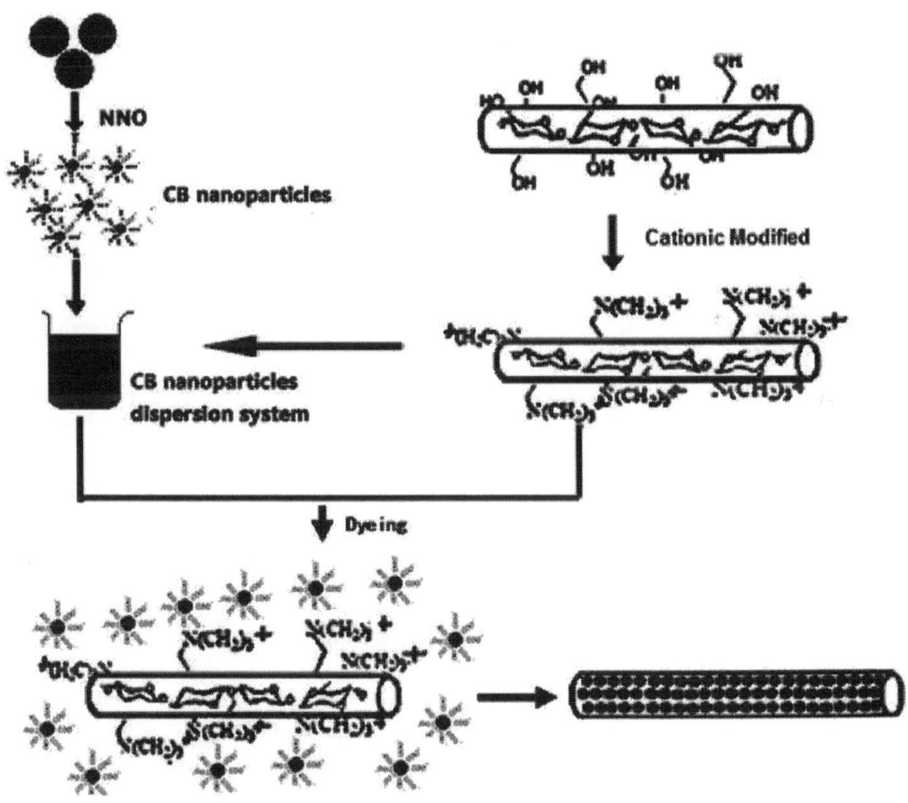

FIGURE 7. The dyeing process of carbon black UMP on modified cotton fabric

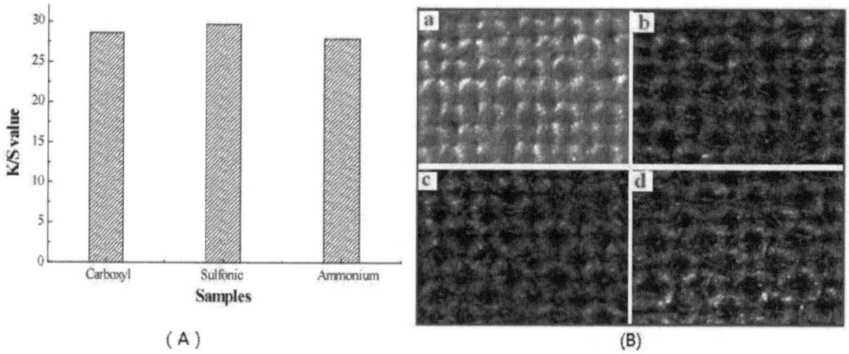

FIGURE 8. Color assessment of CB dyed cellulose matrixes: (A)K/S values; (B) Video microscope microscope:(a) Cotton fabrics without modification, (b) Carboxyl modified cotton fabric deposited with positive NIMs and carbon black UMPs, (c) Sulfonated modified cotton fabric deposited with positive NIMs and carbon black UMPs, (d) Ammonium modified cotton fabric deposited with carbon black UMPs

The results of the K/S presented that the sulfonated cellulose had the higher K/S value of 29.7 than the carboxyl of 28.6 and aminocellulose of 27.9 (Figure 8(A)). The results proved the conclusion about the color depth of the decorated samples. The images (Figure 8(B)) of the samples before and after the dyeing showed that the color depths of the dyed carboxyl and ammonium fabrics were not as deep as the sulfonated cellulose fabric, which further confirmed the color depth conclusion. Especially for the image of the ammonium sample, the fabric surface was uncoated with carbon black UMP, and there were some starred white flake on the fabric.

4.5. DYEING COTTON FABRIC WITH CATIONIC UMP

Surface treatments are effective for UMP because its surface contain polar or polarized functional groups, which can serve as adsorption sites for the hydrophilic or lipophilic groups of the surfactants [1]. The cationic UMP dispersion has great prospects in cellulose dyeing, ink-jet printing, and so on. Compared with the anionic and non-ionic UMP, the cationic UMP has stronger combining force with cellulose substrate, better K/S value, and vividness, thus it can increase the uptake of the UMP, reduce the environment pollution and improve the quality of final products.

For the specific structure, there are a lot of the characteristics on Gemini cationic dispersant. The Zeta potential of the Gemini cationic UMP is higher than that of the ordinary cationic UMP. Therefore, it is easier to color. The dry and wet rubbing fastness is good with both Gemini cationic UMPs after the film fixation (Table 2). With the dispersant dodecyltrimethylammonium, the dyeing of Gemini cationic UMPs exhibits deeper K/S value and the fabric is easier to be colored. The adsorption capacity of the Gemini cationic UMPs to cotton fabric is stronger, and can cover more fiber surface. Cotton fabric shows negative in a water solution, which has a weak charge bonding force with cationic UMPs. The more cationic UMPs containing electropositive, the easier the cotton fabric adsorbing the UMPs [16].

TABLE 2. K/S value and fastness of Gemini cationic UMP dispersion on cotton

Sample	K/S value	Dye rubbing fastness	Wet rubbing fastness
Traditional pigment	42.64	3-4	2
UMP	51.99	4	2

4.6. Dyeing Acrylic Yarns With Cationic Ump

Acrylic fibers can be dyed by cationic UMPs, and the electrostatic attraction between UMPs and acrylic fibers is beneficial to the dyeing properties, especially its light fastness property. Compared with anionic and nonionic UMP systems, the cationic UMPs could bond with the acrylic fibers by electrostatic attraction. Therefore, the UMP presented a better dyeing result and a higher utilization ratio [57].

The dye bath pH presented great influences on cationic UMP dispersion stability during dyeing acrylic fiber with cationic UMP. The increase of UMP dosage resulted in lower dye-uptake and more brilliant colors, and dyeing acrylic yarns with UMP (40%, o.w.f) could reach the maximum K/S value. The K/S value of acrylic yarns reduced after adding adhesives, compared to the cationic dyes, the light fastness was improved significantly [58].

From Figure 9, when the temperature was lower than 65 °C, the dyeing rate was quite low, and the adsorption of pigments was unstable, while the temperature was higher than 65 °C, the dyeing rate increased significantly with the dyeing time prolonging. A higher dyeing temperature would shorten the dyeing time. As the temperature was higher than 95 °C, the dyeing time was only 15 min. The reason was that in a low temperature condition, the acrylic molecular chains were not moving and the adsorption only took place in the surface of fiber. The pigments adsorbed to the yarns were easily re-dissolved into the dye bath.

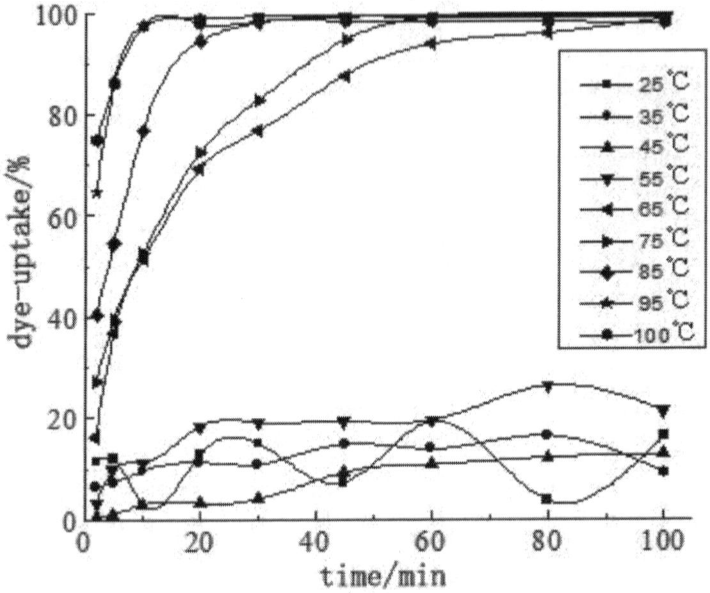

FIGURE 9. Effects of dyeing temperature on dye-uptake of acrylic yarns

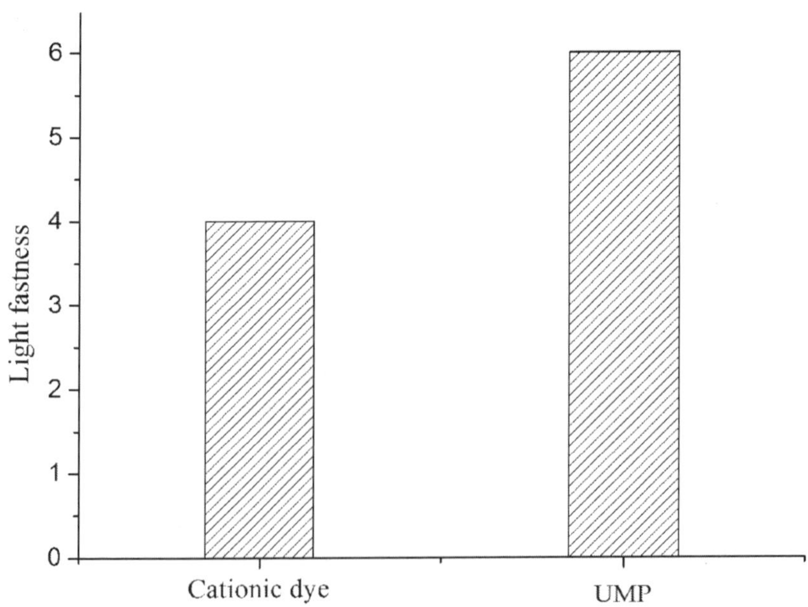

FIGURE 11. Light fastness of acrylic yarns dyed by cationic dye and ultra-fine pigment

In order to evaluate the light fastness of acrylic yarns dyed by cationic dye, the light fastness of cationic dye is attempted and assessed (Figure 10).

Cationic UMP presented better light fastness than that of the cationic dye. Acrylic yarns dyed by cationic UMP presented higher stability to light than that dyed by cationic dye. The main reason was that the chromophoric group of pigment was more difficult to destroy for its stable chemical structure.

5. CONCLUSION

A stable UMP dispersion is able to be prepared via grinding, ultrasonic wave, microencapsulation and microfluidics processes with different dispersants like low molecular weight dispersant, polymeric dispersant, copolymer dispersant and siloxane dispersant. It's more efficient to prepare the uniform and stable UMP dispersion via comprehensive utilization of these dispersing processes.

The key dispersion characteristics of UMP dispersion be evaluated by particle size, disperse stability, Zeta potential, color properties and dyeing properties, which are greatly affected by ionic strength, dispersants and dispersing methods. The disperse stability and color performance are promoted with the smaller particle size and higher charged UMP.

The dyeing properties of UMP on many fabrics such as cotton, silk and wool pretreated with cationic reagent is enhanced. The K/S value and rubbing fastness are able to be distinctly improved. Cationic cotton can also enhance the dyeing properties of aqueous carbon black UMP dispersions due to the higher affinity between carbon black UMP and cationic cotton. The cotton fabric dyed with cationic UMP can also promote the K/S value and color fastness. For acrylic yarns, the increase of UMP dosage results in lower dye-uptakes and more brilliant colors, and dyeing with UMP (40%, o.w.f) can reach the maximum K/S value on acrylic yarns.

As an advanced pigment dyeing method, the UMP takes many advantages of pigment. The specific characteristics of the UMP, including stability and high K/S value, reveal its significant potential on fabric dyeing, and especially for its environment-friendly characteristics, the development of UMP will present a more sharply speed than that of the dyes. Base on the textile application, the further research interests about UMP are focus on the novel dispersing process to obtain stable UMP system, modification with functional organic or inorganic materials, developing the multifunction and multipurpose of UMP.

ACKNOWLEDGEMENTS

The authors are grateful for the financial support of the National Natural Science Foundation of China (21174055) and the 333 Talent Project Foundation of Jiangsu Province (BRA2011184).

REFERENCES

1. T. Vernardakis, Dispersion. Pigment, 2006In: Tracton A A, editor. Coatings Technology Handbook Third Edition. Taylor & Francis Group, LLC2006. 7671

2. J. Gage, 1999Color and Culture: Practice and Meaning from Antiquity to Abstraction.University of California Press, 1218

3. Velmurugan, P., Kim, M., Park, J., Karthikeyan, K., Lakshmanaperumalsamy, P., Lee, K., Park, Y., & Oh, B.2010Dyeing of Cotton Yarn with Five Water Soluble Fungal Pigments Obtained from Five Fungi[J].Fiber Polym, 114598605

4. El -Shishtawy, R, & Nassar, S. Cationic pretreatment of cotton fabric for anionic dye and pigment printing with better fastness properties[J]. Color Technol, 2002

5. K. Hayashi, H. Morii, K. Iwasaki, S. Horie, N. Horiishi, K. Ichimura, 2007Uniformed nano-downsizing of organic pigments through core-shell structuring[J]. Journal of Materials Chemistry, 176527530

6. Kelley, A., Alessi, P., Fornalik, J., Minter, J., Bessey, P., Garno, J., & Royster, T.2010Investigation and Application of Nanoparticle Dispersions of Pigment Yellow 185 using Organic Solvents[J]. ACS Appl Mater Interfaces, 216168

7. Y. Kobayashi, H. Chiba, K. Mizusawa, N. Suzuki, J. M. Cerda-Reverter, A. Takahashi, 2011Pigment-dispersing activities and cortisol-releasing activities of melanocortins and their receptors in xanthophores and head kidneys of the goldfish Carassius auratus[J]. Gen Comp Endocr, 1733438446

8. Lan, Y., & Lin, J.2011Clay-assisted dispersion of organic pigments in water[J]. Dyes Pigment, 9012127

9. Ricaurte-Avella, G., Osorio, Velez. J., Nino, Cardona. J., Correa, J., & Alberto, Rodriguez. H.2010Evaluation of the pigment dispersion in

acrylic resins for the manufacture of artificial teeth[J]. Revista Facultad De Ingenieria-Universidad De Antioquia, 53 5463

10. Ahmed, N., & El -Shishtawy, R.2010The use of new technologies in coloration of textile fibers[J]. Journal of Materials Science, 45511431153

11. Wang, C., Zhang, X., Lv, F., & Peng, L.2012Using carbon black nanoparticles to dye the cationic-modified cotton fabrics[J]. J Appl Polym Sci, 124651945199

12. Y. Cai, M. Pailthorpe, S. David, 1999A new method for improving the dyeability of cotton with reactive dyes[J]. Textile Research Journal, 696440446

13. Wang, C., & Zhang, Y.2007Effect of cationic pretreatment on modified pigment printing of cotton[J]. Materials Research Innovations, 1112730

14. Ishwarlal, R.2000Dyeing anionic cellulosic fabric materials with pigment colors having a net cationic charge using a padding process. US 27

15. T. Koyanagi, 2002Ink sets for cellulose fiber-based recording media, printing method therewith and pinted products therefrom. 23

16. C. Wang, Y. Yin, K. Fang, 2008Investigation of gemini cationic water based pigment dispersion[J]. Materials Research Innovations, 121711

17. C. Diaz-Blanco, Gonzalez. M. Diaz, Monmany. J. M. Daga, T. Tzanov, 2009Dyeing properties, synthesis, isolation and characterization of an in situ generated phenolic pigment, covalently bound to cotton Enzyme and Microbial Technology380 EOF385 EOF

18. J. Dong, S. Chen, D. S. Corti, E. I. Franses, Y. Zhao, H. T. Ng, E. Hanson, 2011Effect of Triton X-100 on the stability of aqueous dispersions of copper phthalocyanine pigment nanoparticles[J]. J Colloid Interface Sci, 36213341

19. K. Huhnerfuss, Bohlen. A. von, D. Kurth, 2006Characterization of pigments and colors used in ancient Egyptian boat models[J]. Spectrochimica Acta Part B-Atomic Spectroscopy, 6110-11 12241228

20. S. Leelajariyakul, H. Noguchi, S. Kiatkamjornwong, 2008Surface-modified and micro-encapsulated pigmented inks for ink jet printing on textile fabrics[J]. Progress in Organic Coatings, 622145161

21. K. Fang, X. Jiang, C. Wang, M. Wu, Y. Yan, 2008Properties of the Nanoscale Hydrophilic Cationic Pigment Based on Quaternary Surfactant[J]. J Disper Sci Technol, 2915257

22. H. Wu, G. Gao, Y. Zhang, S. Guo, 2012Coating organic pigment particles with hydrous alumina through direct precipitation[J]. Dyes and Pigments, 921548553

23. Chang, C., Chang, S., Tsou, S., Chen, S., Wu, F., & Hsu, M.2003Effects of polymeric dispersants and surfactants on the dispersing stability and high-speed-jetting properties of aqueous-pigment-based ink-jet inks[J]. J Polym Sci Pt B-Polym Phys, 411619091920

24. C. Wang, Y. Yin, X. Wang, G. Bu, 2008Improving the Color Yield of Ultra-fine Pigment Printing on Cotton Fabric[J]. AATCC Review 8124145

25. C. Yoon, J. Choi, 2008Syntheses of polymeric dispersants for pigmented ink-jet inks[J]. Coloration Technology, 1246355363

26. N. Bulychev, E. Kisterev, Y. Ioni, O. Confortini, Prez. F. Du, V. Zubov, C. Eisenbach, 2011Suffache Modification in Aqueous Dispersions with Thermo-Responsive Poly (Methylvinylether) Copolymers in Combination with Ultrasonic Treatment[J]. Chemistry & Chemical Technology, 515965

27. S. Fu, C. Du, K. Zhang, C. Wang, 2011Colloidal properties of copolymer-encapsulated and surface-modified pigment dispersion and its application in inkjet printing inks[J]. J Appl Polym Sci, 1191371376

28. Y. M. Chen, R. S. Hsu, H. C. Lin, S. J. Chang, S. C. Chen, J. J. Lin, 2009Synthesis of acrylic copolymers consisting of multiple amine pendants for dispersing pigment[J]. J Colloid Interface Sci, 33414249

29. W. Zhang, S. Deodhar, D. G. Yao, 2010Processing Properties of Polypropylene With a Minor Addition of Silicone Oil[J]. Polymer Engineering and Science, 50713401349

30. J. Wu, L. M. Wang, P. Zhao, F. Wang, G. Wang, 2008A new type of quaternary ammonium salt containing siloxane group and used as favorable dispersant in the surface treatment of CI pigment red 170[J]. Progress in Organic Coatings, 632189194

31. U. Weber, 2010The Effect of Grinding Media Performance on Milling a Water-Based Color Pigment[J]. Chem Eng Technol, 33914561463

32. De Vallejuelo, S., Barrena, A., Arana, G., De Diego, A., & Madariaga, J.2009Ultrasound energy focused in a glass probe: An approach to the simultaneous and fast extraction of trace elements from sediments[J]. Talanta, 802434439

33. X. Jia, C. Caroli, B. Velicky, 1999Ultrasound propagation in externally stressed granular media[J]. Physical Review Letters, 82918631866

34. Johnson, J., Barbato, M., Hopkins, S., & O'Malley, M.2003Dispersion and film properties of carbon nanofiber pigmented conductive coatings[J]. Progress in Organic Coatings, 473198206

35. Y. Yin, C. Wang, Y. Wang, 2012Preparation and colloidal dispersion behaviors of silica sol doped with organic pigment[J]. Journal of Sol-Gel Science and Technology, 62217

36. Fu, S., Xu, C., Du, C., Tian, A., & Zhang, M.2011Encapsulation of CI Pigment blue 15:3 using a polymerizable dispersant via emulsion polymerization[J]. Colloid Surf A-Physicochem Eng Asp, 3811-3 6874

37. Fu, S., & Fang, K.2008Properties of waterborne nanoscale pigment red 122 dispersion prepared by phase separation method[J]. J Appl Polym Sci, 108639683972

38. Squires, T., & Quake, S.2005Microfluidics: Fluid physics at the nanoliter scale[J]. Reviews of modern physics, 773977

39. J. Merrington, P. Hodge, S. Yeates, 2006A High-Throughput Method for Determining the Stability of Pigment Dispersions[J]. Macromolecular Rapid Communications, 2711835840

40. Chang, C., Chang, S., Shih, K., & Pan, F.2005Improving mechanical properties and chemical resistance of ink-jet printer color filter by using diblock polymeric dispersants[J]. J Polym Sci Pt B-Polym Phys, 432233373353

41. F. Nsib, N. Ayed, Y. Chevalier, 2007Selection of dispersants for the dispersion of CI Pigment Violet 23 in organic medium[J]. Dyes Pigment, 741133140

42. D. Xu, K. Fang, S. Fu, 2009Effects of Ethanol on the Stability of Pigment Colloidal Dispersion[J]. Journal of Dispersion Science and Technology, 304510513

43. Wu, H., Huang, K., & Lee, K.2011Supercritical fluid-assisted dispersion of CI pigment violet 23 in an organic medium[J]. Powder Technology, 2063322326

44. P. Karrer, Improvements in or relating to the treatment of cotton fibers preparatory to dyeing. GB Patent 2498421926.

45. Guthrie, J.1947Introduction of Amino Groups into Cotton Fabric by Use of 2-Aminoethylsulfuric Acid [J]. Textile Research Journal, 1711625629

46. H. Wang, D. Lewis, 2002Chemical modification of cotton to improve fibre dyeability[J]. Coloration Technology, 1184159168

47. S. Burkinshaw, X. Lei, D. Lewis, 1989Modification of cotton to improve its dyeability. Part 1-pretreating cotton with reactive polyamide-epichlorohydrin resin[JJournal of the Society of Dyers and Colourists10511391398

48. Hauser, P., & Tabba, A.2001Improving the environmental and economic aspects of cotton dyeing using a cationised cotton[J]. Color Technol, 1175282288

49. H. Jin, 2009The Synthesis and Application of Cationic Polymer (PDA) on Fibers Modification[D]. Master thesis. Jiangnan University;

50. Dubas, S., Chutchawatkulchai, E., Egkasit, S., Lamsamai, C., & Potiyaraj, P.2007Deposition of polyelectrolyte multilayers to improve the color fastness of silk[J]. Textile Research Journal, 776437441

51. J. Wei, 2003Studies on Weighting Methods and Properties of Silk Fiber[D]. Master thesis. Suzhou University;

52. Qin, C., Chen, D., Tang, R., Huang, Y., Wang, X., & Chen, G.2010Studies on silk fabric dyed with a hemicyanine dye[J]. Coloration Technology, 1265303307

53. C. Wang, K. Fang, W. Ji, 2007Superfine pigment dyeing of silk fabric by exhaust process[J]. Fiber Polym, 82225229

54. Murphy, A., John, P., & Kaplan, D.2008Modification of silk fibroin using diazonium coupling chemistry and the effects on hMSC proliferation and differentiation[J]. Biomaterials, 291928292838

55. C. Wang, W. Ji, 2006Dyeing property of the modified wool fabric with superfine pigment[J]. Journal of Textile Research, 2767477

56. X. Yuan, K. J. Fang, 2006Preparation of Waterborne Nanoscale Carbon Black Dispersion with Sodium Carboxymethyl Cellulose[J]. Journal Of Donghua University(English Edition) 235119121

57. R. Yang, M. Dai, 1996Study on Properties of Acid-Modified Multicomponent Copolyesterification Fibre[J]. Journal Of Textile Research, 1712731

58. Y. Zhou, D. Yu, P. Xi, S. L. Chen, 2003Influence of styrene-maleic anhydride copolymers on the stability of quinacridone red pigment suspensions[J]. Journal of Dispersion Science and Technology, 245731737

Chapter 7

Dyeing of Textiles with Natural Dyes

Ashis Kumar Samanta and Adwaita Konar

Department of Jute and Fibre Technology, Institute of Jute Technology, University of Calcutta, India

INTRODUCTION

Textile materials (natural and synthetic) used to be coloured for value addition, look and desire of the customers. Anciently, this purpose of colouring textile was initiated using colours of natural source, untill synthetic colours/dyes were invented and commercialized. For ready availability of pure synthetic dyes of different types/classes and its cost advantages, most of textile dyers/ manufacturers shifted towards use of synthetic colourant. Almost all the synthetic colourants being synthesized from petrochemical sources through hazardous chemical processes poses threat towards its eco-friendliness.

Hence, worldwide, growing consciousness about organic value of eco-friendly products has generated renewed interest of consumers towards use of textiles (preferably natural fibre product) dyed with eco-friendly natural dyes. Natural dyes are known for their use in colouring of food substrate, leather as well as natural fibres like wool, silk and cotton as major areas of application since pre-historic times. Although this ancient art of dyeing textiles with natural dyes withstood the ravages of time, but due to the wide availability of synthetic dyes at an economical price, a rapid decline in natural dyeing continued. However, even after a century, the uses of natural dyes never erode completely and they are being still used in different places of the world. Thus, natural dyeing of different textiles and leathers has been continued mainly in the decentralized sector for specialty products besides the use of synthetic dyes in the large scale sector for general textiles/apparels.

Recently, most of the commercial dyers and textile export houses have started re-looking to the maximum possibilities of using natural dyes for dyeing and printing of different textiles for targeting niche market. Natural dyes produce very uncommon, soothing and soft shades as compared to synthetic dyes. On the other hand, synthetic dyes, which are widely available at an economical price and produce a wide variety of colours, sometimes causes skin allergy and other harmfulness to human body, produces toxicity/chemical hazards during its synthesis, releases undesirable/hazardous/toxic chemicals etc.

For successful commercial use of natural dyes for any particular fibres, the appropriate and standardized techniques for dyeing for that particular fibre-natural dye system need to be adopted. Therefore to obtain newer shade with acceptable colour fastness behaviour and reproducible colour yield, appropriate scientific dyeing techniques/procedures are to be derived. Thus, relevant scientific studies and its output on standardization of dyeing methods, dyeing process variables, dyeing kinetics and test of compatibility of selective natural dyes have become very important, however the information on which is insufficient. That is why, this chapter is very much relevant to the current need of the textile dyers. An attempt has been made here to give scientific overview on dyeing of textiles with natural dyes and related issues.

DEFINITION OF NATURAL DYES/COLOUANTS

The word 'natural dye' covers all the dyes derived from the natural sources like plants, animal and minerals. Natural dyes are mostly non-substantive and must be applied on textiles by the help of mordants, usually a metallic salt, having an affinity for both the colouring matter and the fibre. Transition metal ions usually have strong co-ordinating power and/or capable of forming week to medium attraction/interaction forces and thus can act as bridging material to create substantivity of natural dyes/colourants when a textile material being impregnated with such metallic salt (i.e. mordanted) is subjected to dyeing with different natural dyes, usually having some mordantable groups facilitating fixation of such dye/colourant. These metallic mordants after combining with dye in the fibre, it forms an insoluble precipitate or lake and thus both the dye and mordant get fixed to become wash fast to a reasonable level.

ADVANTAGES AND DISADVANTAGES OF NATURAL DYES/ COLOUANTS

In the recent years, there has been a trend to revive the art of natural dyeing. This is mainly because in some aspects natural colourants are advantageous against synthetic dyes. Some of these advantages along with some limitations (disadvantages) are listed below:

Advantages of natural dyes/ colouants

i. The shades produced by natural dyes/colourants are usually soft, lustrous and soothing to the human eye.

ii. Natural dyestuff can produce a wide range of colours by mix and match system. A small variation in the dyeing technique or the use of different mordants with the same dye (polygenetic type natural dye) can shift the colours to a wide range or create totally new colours, which are not easily possible with synthetic dyestuffs.

iii. Natural dyestuffs produce rare colour ideas and are automatically harmonizing.

iv. Unlike non-renewable basic raw materials for synthetic dyes, the natural dyes are usually renewable, being agro-renewable/vegetable based and at the same time biodegradable.

v. In some cases like harda, indigo *etc.,* the waste in the process becomes an ideal fertilizer for use in agricultural fields. Therefore, no disposal problem of this natural waste.

vi. Many plants thrive on wastelands. Thus, wasteland utilization is an added merit of the natural dyes. Dyes like madder grow as host in tea gardens. So there is no additional cost or effort required to grow it.

vii. This is a labour intensive industry, thereby providing job opportunities for all those engaged in cultivation, extraction and application of these dyes on textile/food/leather etc.

viii. Application of natural dyes has potential to earn carbon credit by reducing consumption of fossil fuel (petroleum) based synthetic dyes.

ix. Some of its constituents are anti-allergens, hence prove safe for skin contact and are mostly non-hazardous to human health.

x. Some of the natural dyes are enhanced with age, while synthetic dyes fade with time.

xi. Natural dyes bleed but do not stain other fabrics, turmeric being an exception.

xii. Natural dyes are usually moth proof and can replace synthetic dyes in kids garments and food-stuffs for safety.

Despite these advantages, natural dyes do carry some inherent disadvantages, which are responsible for the decline of this ancient art of dyeing textiles.

Limitation/ disadvantages of natural dyes/ colouants

i. It is difficult to reproduce shades by using natural dyes/colourants, as these agro- products vary from one crop season to another crop season, place to place and species to species, maturity period etc.

ii. It is difficult to standardize a recipe for the use of natural dyes, as the natural dyeing process and its colour development depends not only on colour component but also on materials.

iii. Natural dyeing requires skilled workmanship and is therefore expensive. Low colour yield of source natural dyes thus necessitates the use of more dyestuffs, larger dyeing time and excess cost for mordants and mordanting.

iv. Scientific backup of a large part of the science involved in natural dyeing is still need to be explored.

v. Lack of availability of precise technical knowledge on extraction and dyeing techniques.

vi. The dyed textile may change colour when exposed to the sun, sweat and air.

vii. Nearly all-natural dyes with a few exceptions require the use of mordants to fix them on to the textile substrate. While dyeing, a substantial portion of the mordant remains unexhausted in the residual dye bath and may pose serious effluent disposal problem.

viii. With a few exceptions, most of the natural dyes are fugitive even when applied in conjunction with a mordant. Therefore, sometimes their colour fastness performance ratings are inadequate for modern textile usage.

CLASSIFICATION OF NATURAL DYES/ COLOUANTS

Natural dyes can be classified (Gulrajani & Gupta, 1992) in a number of ways. The earliest classification was according to alphabetical order or according to the botanical names. Later, it was classified in various ways, e.g. on the basis of hue, chemical constitution, application class etc.

a. In "treatise on permanent colours" by Bancroft, natural dyes are classified into two groups: 'Substantive Dyes' such as indigo, turmeric etc. which dye the fibers directly and 'Adjective Dyes' such as logwood, madder etc. which are mordanted with a metallic salt.

b. Humme classify the colouring matter as 'Monogenetic Dyes', those produce only one colour irrespective of the mordant present on the fibre or applied along with the dye and 'Polygenetic Dyes', those produce different colour with different mordant applied, e.g., alizarin (Dedhia, 1998)

c. In the colour index the natural dyes are classified according to the hue (Predominating colour). The number of dyes in each hue are as follows in table-1:

Table 1. Showing the number of natural dyes in each hue as per the colour index.

CI Natural	No. of Dyes	Percent
Yellow	28	30.4
Orange	6	6.5
Red	32	34.8
Blue	3	3.3
Green	5	5.5
Brown	12	13
Black	6	6.5

On the basis of hues, natural dyes can be classified as follows:

i. Red colour dyes: most red dyes are hidden in roots or barks of plants or camouflaged in the bodies of dull grey insects. They are almost invariably based on anthraquinone and its derivatives. These dyes are stable to light and washing.

ii. Yellow colour dyes: Yellow is the liveliest and perhaps the most abundant of all hues in nature. About 90% of the yellow dyes are flavonoids. Generally, they produce pale shade with quicker fading except turmeric, which produce dull deep shade but considered to be susceptible to light as they emit fluorescence. Wash fastness rating of natural yellow dyes ranges from fair to excellent, e.g., tesu, turmeric, kapila.

iii. Blue colour dyes are indigo and woad, give excellent fastness to light and washing.

iv. Black colour dyes: Black shades, generally obtained from tannin rich plant natural dyes and appreciably substantive towards cellulosic and protein fibre, imparts good overall fastness properties. Examples – logwood, harda, custard apple etc.

a. On the basis of origin, natural dyes are broadly classified into three categories: vegetable, mineral and animal origin. About 500 vegetable origin dyes, colouring matter derived from root, leaf, bark, trunk or fruit of plants, are as follows in table-2

Table 2. Showing some common natural dyestuffs obtained from different vegetable origin.

Part of the Plants	Dyestuffs
Root	Turmeric, Madder (Manjistha), Onions, Beet-root
Bark/ Branches	Purple bark, Sappan wood, Shillicorai, Khair, Red, Sandalwood
Leaf	Indigo, Henna, Eucalyptus, Tea, Cardamon, Coral Jasmine, Lemon Grass
Flowers (Petals)	Marigold, Dahlia, Tesu, Kusum
Fruits/Seeds	Latkan, Pomegranate rind, Beetle nut, Myrobolan (Harda)

Mineral origin colourants are derived from specific mineral natural source or so-called mineral colours are produced from purified inorganic compounds. Some of the important mineral colourants are chrome-yellow, iron-buff, narkin-yellow, Prussian-blue and manganese brown. Animal origin lac, cochineal and kermes have been the principal natural dyes yielding from the insects.

a.Natural dyes can also be classified on the basis of their chemical constitution (Dedhia, 1998).

i. Indigoid dyes: Indigo and tyrian purple are the most common examples of this class. Another blue dye, woad also possesses indigo as the main dyeing component.

ii. Anthraquinone dyes: Almost all the red natural dyes are based on the anthraquinoid structure having both plant and mineral origin. Madder, lacs, kermes, cochineal are some of the dyes possess this type of structure. These are generally mordant dyes.

iii. Alphanaphthoquinones: Typical example of this class is lawsone (henna), cultivated mainly in India and Egypt. Another similar dye is juglone, obtained from the shells of unripe walnuts. These dyes are generally disperse dyes and give shades of orange.

iv. Flavonoids, which yield yellow dyes can be classified under flavones, isoflavones, aurones and chalcones. Flavones are colourless organic compounds. Most of the natural yellows are derivatives of hydroxyl and methoxy substituted flavones and isoflavones. Common example is weld (containing luteolin pigment) giving brilliant and fast colours on both wool and silk.

v. Di-hydropyrans: Closely related in chemical structure to the flavones are substituted di- hydropyrans, *viz.* haematin and its leuco form, haematoxylin. These are important natural dyes for dark shades on silk, wool and cotton. Logwood, brazil wood and sappan-wood are the common example.

vi. Anthocyanidins: The naturally occurring member of this class includes carajurin, a direct orange dye for wool and cotton. It is obtained from the leaves of bignonia chica.

vii. Carotenoids: The class name carotene is derived from the orange pigment found in carrots. In these, the colour is due to the presence of long conjugated double bonds.

b. Another method of classifying natural dye is on the basis of the method of application (Gulrajani & Gupta, 1992).

i. **Mordant dyes** are dyestuffs which require a mordant in their application as they have no affinity for the fiber being dyed. A mordant dye should have electron donating groups capable of forming a complex with the transition metal salt, e.g., madder, fustic, persian, berries, kermes, cochineal etc.

ii. **Vat dyes** are water insoluble dyes which are first converted to their water soluble form (reducing with Na- hydrosulphite and then solubilising it with alkali) and then applied to the fibres. The true colour is produced only on oxidation followed by treatment with a hot soap solution, *e.g.*, indigo.

iii. **Direct dyes** are those dyes that have tremendous affinity for the cellulosic fibres. They are dyed from a boiling dye bath. Turmeric, harda, pomegranate rind etc. are the few of the direct natural dyes.

iv. **Acid dyes** are applied from an acidic medium. The dye molecules have either sulphonic or carboxylic group (s) which can form an electrovalent bond with amino groups of wool and silk. An after treatment with tannic acid known as back tanning improves the fastness of these type of dyes, *e.g.,* saffron.

v. **Disperse dye** has a relatively low molecular mass, low solubility and no strong solubilizing groups. Disperse dyes can be applied on to hydrophobic synthetic fibre from neutral to mildly acidic pH. They can also be applied to silk and wool. These dyes can be post-mordanted with chromium, copper and tin salts, *e.g.,* lawsone and many other flavone and anthroquinone dyes.

vi. **Basic or cationic dyes** on ionization give coloured cations and form an electrovalent bond with the –COOH group of wool and silk. These dyes are applied from neutral to mildly acidic pH. These dyes have poor light fastness, *e.g.,* berberine.

Extraction process of colour component from natural dyes

Extraction of colour component from source natural dye material is important step for dyeing any textile substrate to maximize the colour yield. Moreover, standardization of extraction process and optimizing the extraction variables both, for a particular source natural dye material have technical and commercial importance on colour yield and cost of extraction process as well as dyeing cost. The natural dyes can be taken from various vegetable sources like flowers, stem or wood, roots, bark, etc. as well as animal sources and mineral sources. The colour component present in these sources needs to be extracted so that it can be applied suitably on textiles. Natural dyes of different origin can be extracted using aqueous method i.e. by using water for the extraction with or without addition of salt/acid/alkali/alcohol in the extraction bath, supercritical fluid extraction, enzyme assisted extraction, alcoholic/organic solvent extraction by using relevant extracting equipment or soxhlet extraction method with use of alcohol and benzene mixture and finally to filterate, evaporate and to dry using ultra filtration equipment or centrifuge rotatory vacuum pump/or by extraction under reduced pressure. Now a days, there has been industrial methods available for extracting colour components/purified colour substances from natural dyes for their easy applications.

The collected source material is generally shadow dried in air or sun dried within a temperature range of 37-40°C for the moisture content of the source natural dye material is reduced to 10-15% with proper drying since most of the material have moisture content of 40-80% and can not be stored without drying. After drying, grinding is carried out to break down the material into very small units or preferably powder form. Extraction refers to separating the desired colour component by physical or chemical means with the aid of a solvent. Optimum conditions of extraction variables are determined through extracting the natural colour component from source material by varying extraction parameters of liquor and measuring the optical density of corresponding coloured liquor by using spectrophotometer. Also, the gravimetric yield of colour can be measured by filtering the extraction liquor through standard filtration process followed by evaporation of solvent, washing and finally drying to get the purified natural colour.

Aqueous extraction system

For optimizing the extraction method of colour component in aqueous medium, dried and finely cut source material of natural dye is grinded in powdered form and then the colour component is extracted in water employing a standard process. The aqueous extraction of dye liquor is carried out under varying condition, such as time of extraction, temperature of extraction bath, pH of extraction liquor, concentration of colour-source material (powdered form of source natural dye material) and Material-to-liquor ratio (MLR). In each case, the optical density or absorbance value at a particular (maximum) absorbance wavelength for the aqueous extract of the natural dye material can be estimated using UV-Vis absorbance spectrophotometer.

Many scientists have reported the optimized process of extraction of natural dyes from source. Colour from leaves of eucalyptus hybrid, seeds of cassia tora and grewia optiva are extracted by using aqueous medium under varying conditions (Dayal & Dobhal, 2001). Natural dyes are extracted from biomass products namely cutch, ratanjot, madder (Khan et al, 2006) and from hinjal, jujube bark (Maulik & Pradhan, 2005) in aqueous medium. An attempt has been made to extract natural dye from the coffee-seed for its application in dyeing textiles like cotton and silk (Teli & Paul, 2006). Grey jute fabric is dyed with extracts from deodar leaf (Pan et al,

2003) jackfruit wood and eucalyptus leaf by soaking it soft water and boiling it for 4 hours separately. Extraction (Verma & Gupta, 1995) of natural dyes is also reported from overnight soaked wattle bark in distilled water followed by boiled it in pressure vessel and filtered it to obtain a residual dye powder of about 15 to 20 % (w/w) of the bark. Colours are extracted from marie gold and chrysanthemum flowers by boiling the dry petals with acidified or salt water and reported it to be the best (Deo & Paul, 2000; Sarkar et al 2005 & 2006; Saxena et al, 2001). Natural colour extraction process has also been optimized in aqueous media for various source natural dye materials as follows (Konar, 2011):

- Pomegranate Rind: Pre-cut and dried rind is initially crushed to powder form and then it is extracted in water using an optimized condition of extractions using MLR- 1:20, temperature -90°C and time - 45 min and then it is filtered to obtain approximately 40% (w/w) clear extract of coloured aqueous solution of pomegranate rind having pH 11.

- Mariegold (Genda): Dried petal of mariegold is initially crushed to powder form and then extracted in water using an optimized condition of extraction using MLR 1:20 at 80°C for 45 min at pH 11 and then it is filtered to obtain approximately 40% (w/w) coloured aqueous extract of mariegold.

- Babool (Babla): Sun- dried chips (pre-cut) of babool bark is initially crushed to powder form and then it is extracted in water using an optimized condition of extractions, by boiling in water at 100°C for 120min. and using MLR 1:20 and then it is filtered to obtain 40% (w/w) clear extract of coloured aqueous solution of babool having pH 11.

- Catechu (Khayer): Pre-dried powder of catechu is initially crushed to powder form and then extracted in aqueous medium using an optimized condition of extractions by heating in water bath at 90°C having MLR 1:20 and then it is filtered to obtain 40% (w/w) extract of coloured aqueous solution of catechu having pH 12.

- Jack fruit wood: Pre-cut and dried chips of jack fruit wood is initially crushed to powder form and then colour is extracted in water using an optimized conditions of extractions by boiling in water at 100°C for 30 minutes and using ML ratio (MLR) 1:10 and then it is filtered to obtain 40% (w/w) clear extract of coloured aqueous solution of jack fruit wood having pH 11.

- Red sandal wood: Dried pre-cut chips are crushed to powder form and colour is extracted under optimized conditions by heating it in water at 80°C for 90 minutes at pH 4.5 and MLR 1:20.

EXTRACTION BY NON-AQUEOUS AND OTHER SOLVENT ASSISTED SYSTEM

Due to increasingly stringent environmental regulations, supercritical fluid extraction (SFE) has gained wide acceptance in recent years as an alternative to conventional solvent extraction for separation of organic compound in many analytical and industrial process. In recent past decade, SFE has been applied successfully to the extraction of a variety of organic compounds from herbs, other plant material as well as natural colourant from source natural dye material. With increasing public interest in natural products, SFE may become a standard extraction technique for source natural dye material and other herbs and food items. Supercritical fluid extraction using carbon dioxide as a solvent has provided an excellent alternative to the use of chemical solvents. Over the past three decades, supercritical CO_2 has been used for the extraction and isolation of valuable compounds from natural products.

Supercritical fluids are utilized to extract and purify natural colourant from eucalyptus bark (Vankar et al, 2001). Extraction of dye from food is best achieved with ethanol/oxalic acid. The comparative behaviour of other red food dyes is also studied and a process is developed for the extraction of natural dye from the leaves of teak plant is carried out using aqueous methanol (Nanda et al, 2001). A brick red shade from dyeing for silk/wool using the isolated dye in presence of different mordants is achieved. Attempts (Bhattacharya, 2002; Patel & Agarwal, 2001) has been made to standardize colourant derived from arjun bark, babool bark and pomegranate rind. Extraction (Agarwal et al 1992; Singh & Kaur, 2006) of well grounded henna leaves, directly in a solvent assisted dyeing process, employing organic solvent:water (1:9) as the dyeing medium is studied and superior dyeing properties are obtained, when applied to polyester. Natural dye (Raja & Kala, 2005) is obtained from the grape skin waste by using soxhlet extractor, and latter on distilled it under vacuum to obtain the concentrated dye solution. Colourant/dye is extracted by using a reflux condenser; source dye material is refluxed for 1 hour and filtered it to yield natural colourant (Eom et al, 2001).

EXTRACTION BY ACID AND ALKALI ASSISTED SYSTEM

Colour from euphorbia leaves (Dixit & Jahan, 2005) under acidic pH by adding hydrochloric acid and under alkaline pH by adding sodium carbonate both, in aqueous media are extracted for dyeing silk fabric. Extraction of colour in alkali media from nuts of acacia catechu (Sudhakar, 2006) is carried out for colouration of protein fibre based fabric. Dye extracted from jatropha seed gives a range of bright, even and soft colours on textiles when extracted under acid/alkali condition (Radhika & Jacob, 1999). Extraction of colour component from jackfruit wood under various pH conditions is carried out and reported that the optimum conditions for extraction is at pH-11.0 (Samanta et al, 2007). Red colour is extracted from red sandal wood (Samanta et al, 2006)under various pH conditions and it is reported that the optimum conditions for extraction of colour component is acidic pH like

4.0. Orangish yellow colour is extracted from tesu (palash flower) and maroon red colour can be obtained from Indian madder when extracted at alkaline conditions under aqueous medium (Samanta et al, 2010 & 2011).

Natural colour extraction by other methods

For ultrasound assisted extraction process (Sumate et al, 2008; Tiwari et al 2010) of natural colour, a standard extraction protocol may be used as 250 ml of solvent and 25 gm of powdered source natural dye material are taken in 500 ml beaker being immersed into the ultrasonic bath with working frequency of 27-30 MHz at 160 V under a controlled water level at about 2-3 cm from the bottom of the bath. To estimate the extraction yield at different time, temperature, pH and MLR for optimizing the extraction variables, 1.0 ml liquid is pipette out and then diluted to make 10.0 ml volume in each case. This solution is centrifuged at about 2000 rpm to remove the suspension. Finally, the concentration (% w/w) of the diluted solution is measured spectrophotometrically at a definite wavelength (λmax). For enzyme assisted extraction process (Tiwari et al 2010) single or mixed enzyme (e.g. pectinase : cellulase, 2:1) is sprayed on source material and left for overnight for better soaking. This material is then taken into 500 ml conical flask with 250 ml water of pH 10 and shaken in orbital shaker at 150 rpm for 40-80 minutes at optimum temperature. The extraction solution is ready for dyeing textile material or can be filtered and drying as ready purified dye material for further use.

PURIFICATION AND CHARACTERIZATION OF NATURAL DYES

The aqueous extraction of the corresponding dye solution is double filtered in fine mesh nylon cloth and sintered glass crucible and the filtrate is evaporated using a vacuum oven at lower temperature (70°C) to a semi-dried solid mass and the same is then put in a cage of the wrapped filter paper and further subjected to extraction in soxhlet apparatus using 1:1 alcohol:toluene mixture for 10 cycles for 2h at 70ºC. The alcohol- toluene extract of the colour components is finally subjected to evaporation in a water bath at 50ºC to get a semi-dry mass of the pure colour components. Finally, this dry mass of the colour components is washed with 100% acetone followed by washing with methyl alcohol and final drying in air to obtain the dry powder of the pure colour components of the corresponding natural dyes.

For characterization, purified dye powder is to be taken for preparation of 1% aqueous dye solution separately and is subjected to wavelength scan in a micro processor or computer attached UV-Vis absorbance spectrophotometer for 190-1100 nm range. Further, individual purified natural dye powder is washed once again in distilled water and in 100% acetone in sequence before final drying and may be subjected to FTIR Spectroscopy study in double beam FTIR spectrophotometer using KBr disc technique for characterization of its chemical nature and functional group present in the natural dyes, which are responsible for solubilisation and mordanting power of natural dyes as well as its hypsochromic/bathochromic shift of the main hue.

For study of thermal behaviour by DSC (differential scanning calorimetry) study, individual purified natural dye powder is to be washed in distilled water followed by further washing in 100% acetone before final drying and then may be subjected to DSC or TGA (thermogravimetric analyser) study by standard method, for determining the different transition temperature of the purified dyes including temperature of degradation/dissociation. Thermal characterization by DSC/TGA is necessary for understanding the nature of thermal dissociation of natural dye component at different dyeing temperatures as well as application temperature.

UV-Visible spectroscopic study

UV-Vis spectral scan of aqueous/non-aqueous extract/solution of purified natural dyes having both UV-zone and visible zone (190-700 nm or higher) indicating peaks and troughs in different wave length shows its main hue, absorption etc. Peaks and troughs in visible zone thus indicate main colour and absorption. UV-Zone with/without peaks shows the property of the dye under UV-light, this may be correlated with fastness behaviour.

UV-Visible spectroscopic studies are carried out by different scientists (Erica et al, 1995) to identify the UV-Vis spectral scan of a number of natural dyes viz, madder, cochineal, indigo, etc., using different solvents for extraction. Neem bark (Mathur et al, 2003) colourant shows two absorption maxima at 275 and 374 nm while beet sugar shows three absorption bands at 220, 280 and 530 nm as per recent study (Mathur et al, 2001). The visible spectra of ratanjot (Gulrajani et al, 1999) at acidic pH showed maximum absorption around 520-525 nm, but under alkaline pH there is a shift to 570 nm and another peak at 610-615 nm and red sandal wood shows a strong absorption peak at 288 nm , the maximum absorption at 504 and 474 nm at pH 10 in methanol solution (Gulrajani et al, 2003). Gomphrena globosa (Sankar & Vankar, 2005) flower colourant shows one major peak at 533 nm. The dye does not show much difference in the visible spectrum at pH 4 and 7. Absorption for dyes extracted from mimusops elengi and terminalia arjun are reported that depending on the concentrations of dyes in the dye-bath, the dye absorbtion on the fibre varies from 21.94 % to 27.46 % and 5.18

% to 10.78% respectively (Bhuyan et al, 2004). The colour components isolated from most of the barks contain flavonoid moiety. Extraction, spectroscopic and colouring potential studies of the dye in ginger rhizome (zingiber officinale) is studied and reported (Popoola et al, 1995) that the dye is soluble in hydroxyl organic solvents and gives one homogenous component of Rf value of 0.86 on chromatographic separation having wavelength of maximum absorption at 420 nm. Aqueous extract of different source natural dyes including red sandal wood, manjistha, tesu, cutch etc. have been characterized by UV-vis spectra to optimize the extraction conditions.

Chromatographic analysis

Thin layer chromatatography (TLC) is used by many workers to identify natural dyes in textiles (Kharbade et al, 1985). Dyes detected are insect dyes and vegetable dyes viz., yellow, red and blue colours. The natural scale insect, madder and indigoid dyes are also analysed by HPLC (Koren, 1994). TLC chromatography analysis (Guinot et al, 2006) is used to carry out a preliminary evalution of plants containing flavonoids (flavonols, flavones, flavanones, chalcones/ aurones, anthocynanins), hydroxycinnamic acids, tannins and anthraquinones, which are the phylo-compounds (colour compounds) found in the plants. Identification of dyes in historic textiles through chromatographic and spectrophotometric methods as well as by sensitive colour reactions is highlighted (Blanc et al, 2006) and further the retention of carminic acid, indigotin, corcetin, gambogic acid, alizarin flavanoid, anthraquinone and purpurin are also studied (Szostek et al, 2003). A non-destructive method is reported for identifying faded dyes on textiles fabrics through examination of their emission and excitation spectra. The quantitative and qualitative analysis of red dyes such as alizarin, purpurin, carminic acid etc. by HPLC are also investigated/analysed (Balankina et al, 2006). High Performance Liquid Chromatography (HPLC) has been also used by several workers to identify natural dyes.

The separation and identification of natural dyes is carried out from wool fibres using reverse phase HPLC with a C-18 column (Mc Govern et al, 1990). Two quaternary solvent systems and one binary solvent system are used to obtain chromatograms of dyes, isomers and minor products present in the sample. A linear gradient elution method has been applied to the HPLC analysis of plant and scale insect, red anthraquinonoid, mordant dyes, and molluscan blue, red purple and indigoid vat dyes (Koren, 1994). The method enables the use of the same elution programme for the determination of different chemical classes of dyes. In addition, it significantly shortens the retention time of natural anthraquinonoid dyes. Quantitative analysis of weld by HPLC shows that after a 15 min extraction in a methanol-water mixture, 0.45% luteolin, 0.36 % luteolin 7-glucoside and 0.23% luteolin-3'7- diglucoside are obtained (Cristea et al, 2003). HPLC analysis of indigo is reported (Son et al, 2007) and it is found that as the dyeing time is increased, structural changes of indigo component are attributed to decrease in colour strength of dyeing. Detection of annatto dyestuff, norbixin and bixin is reported by means of derivative spectroscopy and high performance liquid chromatography (HPLC) (Bhattacharya, 1999). The sample preparation involved extraction with acetone in the presence of HCl and removal of water by evaporation with ethanol. This residue is dissolved in chloroform-acetic acid for derivative spectroscopy or with acetone for HPLC. Derivative spectra are recorded from 550-400nm. Analysis of cochineal colour in foods utilizing methylation with diazomethane is carried out emplying TLC and HPLC using a mobile phase of butanol/ ethanol and 10% acetic acid.

Test of toxicity, biotechnological processing and environmental impact of natural dyes

Toxicity is the ability of a substance to cause damage to living tissue, impairment of nervous system or severe illness when ingested, inhaled or being absorbed by skin. The toxicity (Zippel, 2004; Joshi & Purwar, 2004) data provide evidence about the adverse effect of natural dyes to human body. The LD50 is the best-known figure for toxicity rating of any substance. It describes the 'lethal dose for 50% of the test animals' which is the amount of substance in kg/kg of body weight which kills half of the animals.

Most of the natural dyes are found to be non-carcinogenic in nature. Moreover, natural dyes have positive effect on antifungal and anti bacterial growth. The crude methanolic extracts of stem and roots stem, leaves, fruit, seeds of artocarpus hetrophyllus (Khan et al, 2003) and their subsequent partitioning with petrol, dichloromethane, ethyl acetate and butanol fractions exhibit a broad spectrum of antibacterial activity. The butanol fractions of the root, bark and fruit are found to be the most active. None of the fraction is found to be active against the fungi tested. Mariegold (http://www.mdidea.com, 2005) shows negative test against microbiology control E-coli and salimonella. Chemo-preventative effects (Dwivedi & Ghazaleh, 1997; Dwivedi & Zhang, 1999; Benencia & Courreges, 1999) of red sandal wood oil are observed on skin papillomas in mice. Further, it is studied for red sandal wood's prevention of skin tumor development in CD1 mice and antiviral activity against herpes simplex virus-1 and 2. The hepatoprotective (Gilani & Janbaz, 1995) activity of an aqueous– methanol extract of rubia cardifolia (madder) is investigated against acetaminophen and CCL4-induced damage. Acetaminophen produced 100% mortality at a dose of 1 g/K in mice while pretreatment of animals with plant extract (natural dye source material) reduced the death rate to 30%. Test of acacia catechu (cutch), kerria lacca (lac), quercus infectoria (gallnut), rubia cordifolia (madder) and rumex maritimus (golden dock) against pathogens like escherichia coli, bacillus subtilis, klebsiella pneumoniae, proteus vulgaris and pseudomonas aeruginosa are also reported (Singh et al, 2005). Minimum inhibitory concentration is found to be varying from 5 to 40 μg. Using a bioassay-directed purification scheme, the active antibacterial principle from caesalpina sappan (sappan wood or red wood) is isolated and identified (Hong & Lee, 2004). The trypan blue dye exclusion test shows that brazilian lacks cytotoxicity against vero cells; it has potential to be developed into an antibiotic. It is reported (Bhattacharya et al, 2004) that arjun bark, babool bark and pomegranate rind are eco-safe, however sometime contains traces amount of red listed heavy chemicals in permissible limit. A critical and realistic evaluation of dyeing with vegetable dyes highlighting its metal toxicity of substances used in processing has been reported (Shenai, 2002) and it is mentionworty that mordanting with metal salt as pre-requisite for application of most of the natural dyes may contaminate the dyed textiles with objectionable heavy metals resulting carcinogenic effect. Therefore, selection of mordanting metal salt and its purity are important criteria to produce eco-friendly natural dyed textiles. Attempts (Mondhe & Rao, 1993a & 1993b) has been made to prepare azo-alkyd dyes by the reduction of nitro alkyds, followed by diazotization of aminoalkyds and coupling with different phenolic compounds present in jatropha curcas seed oil confirmed by using IR spectra.

TYPES OF MORDANTS

Limitation on colour yield and poor fastness properties prompted a search for ideal mordants, the chemicals which increase natural dye uptake by textile fibres. Different types of mordants yield different colours even for the same natural dye. Therefore, final colour, their brilliance and colour fastness properties are not only dependant on the dye itself but are also determined by varying concentration and skillful manipulation of the mordants. Thus, a mordant is more important than the dye itself. Moreover, the ideal mordant for bulk use should produce appreciable colour yield in practicable dyeing conditions at low cost, without seriously affecting physical properties of fibre or fastness properties of the dyes. Also, It should not cause any noxious effect during processing and the dyed textile material should not have any carcinogenic effect during use. Mordants can be classified into the following categories:

Metallic mordants

They are generally metal salts of aluminium, chromium, iron, copper and tin. The metallic mordants are of two types.

Brightening mordants

- **Alum:** Among all types of alum, potash alum is cheap, easily available and safe to use mordant. It usually produces pale versions of the prevailing dye colour in the plant.

- **Chrome (potassium dichromate):** It is also referred to as red chromate. It is relatively more expensive. However, Cr^{3+} or Cr^{6+} is considered to be harmful for human skin as objectionable heavy metal beyond a certain limit of its presence. Its use has been limited as per the norms of the eco-standards. The dichromate solution is light sensitive and therefore it changes colour under light exposure.

- **Tin (stannous chloride)**: It gives brighter colours than any other mordant. However, they are oxidized on exposure to air and may impart a stiff hand to the fabric. Stannous chloride also causes higher loss of fabric tenacity (tensile strength) if applied beyond a certain concentrations.

Dulling mordants

- **Copper (cupric sulphate):** Known as blue vitriol, it is readily soluble in water and easy to apply. It gives some special effects in shades, which otherwise cannot be obtained. However, copper beyond a certain limit is also under the eco-standard norms as objectionable heavy metals.

- **Iron (ferrous sulphate):** It is also known as green vitriol and is readily soluble in water. It is used for darkening /browning and blackening of the colours/ shades. It is easily available and one of the oldest mordants known. It is extensively used to get grey to black shades.

Tannins

The term 'tanning agent' is given initially to those water-soluble cellulosic materials that predicates gelatin from solution. But all gelatin precipitation did not identified as tanning agent. Tannins are polyphenolic compounds having capacity of gelling under certain conditions. (a) It may be hydroysable pyrogallol tannins exemplified by 'tannic acid', by Chinese or Turkish gallotannins (galls) and by Sicilain and Stagshorn sumac, (b) hydroysable ellagitannins that give ellagic acid or similar acids on hydrolysis, exemplified by valonea, chestnut, and (c) condensed or catechol tannins that contain little or no carbohydrates and are converted to acids to insoluble amorphous polymers. Among the tannins, myrobalan (harda) and galls/sumach are most important.

Oils type mordants

Vegetable oils or Turkey red oil (TRO) are such type of mordants. TRO as mordant is mainly used in the dyeing of deep red colour from madder. The main function of the TRO as oil mordant is to form a complex with alum when used as a main mordant. Sulphonated oil posses better binding-capacity than the natural oils. Oil mordanted samples exhibit superior fastness and hue.

DIFFERENT MORDANTING METHODS AND APPLICATION OF NATURAL DYES

Mordanting can be achieved by pre-mordanting (before dyeing), simultaneously mordanting and dyeing or it may be a post mordanting system (after dyeing). Different types of mordants can be applied on the textile to increase the dye uptake of natural dyes. Extensive work has been reported (Paliwal, 2001; Jahan P & S, 2000; Sengupta, 2001; Prabu & Premraj, 2001; Sunita & Mahale, 2002; Moses, 2002; Rani & Singh, 2002; Bain et al, 2002; Paul et al, 2002) for dyeing of textiles with natural dyes adopting specific mordanting system for a particular textile material.

In pre-mordanting method, the textile substrate is first treated in aqueous solution of mordant for optimized time (e.g. 30 - 60 minutes) and temperature (e.g. 70 – 100 °C) with a ML ratio of 1:5 to 1:20 and then dried with or without washing. The mordanted textile material is then dyed following optimized dyeing conditions may be required as salt, soda ash or acid depending on type of textile material and type of natural dye. After dyeing, the textile material is washed properly and soaping is carried out by 2 g/L industrial soap solution as described in standard method of AATCC or ISO method.

For simultaneous mordanting and dyeing system, the textile substrate is immersed in a dye bath solution containing both mordant and dye in a definite quantity and dyeing may be started at the pre-determined optimum condition. Dyeing auxiliaries may be added as required for the standard dyeing process. However, for optimization of dyeing condition, dyeing process variables can be studied for specific fibre-mordant-natural dye system in order to maximize colour yield on textiles. After dyeing, the textile material is washed properly and soaping is carried out by 2 g/L industrial soap solution.

In case of post-mordanting method of natural dyeing, the dyeing process is carried out for bleached textiles in the absence of mordant at pre-determined dyeing condition and the dyed fabric is treated in a separate bath called saturator containing suitable mordanting solution. Treatment condition may vary depending on type of fibre, dye and mordant system. After dyeing, the textile material is washed properly and soaping is carried out by 2g/L industrial soap solution.

There is study (Dayal et al, 2006) for effect of copper sulphate and potassium dichromate on silk, wool and cotton fibre and reported their effects on colour fastness properties. The wool treated with metal ions such as Al(III), Cr (VI), Cu (II), Fe (II), Sn (II) and rare earths such as La (III), Sm (III) are used for beet sugar colourant, it can withstand the requirement of BIS fastness standards. Optimization (Agarwal et al, 1993) of the various concentrations of mordant are reported for shades can be produced by 0.15% of alum, 0.08% copper sulphate and stannous chloride, 0.04% ferrous sulphate and 0.06% potassium dichromate on mulberry silk fabric. Extraction of natural dye from the leaves of teak plant by using aqueous methonal produced brick red shade on dyeing of silk/ wool using the isolated dye in presence of different mordants as it is reported (Nanda et al, 2001). Irrespective of mordanting methods, silk (Mahale et al, 2003) treated with potash alum shows increase in colour when subject to sunlight test and those treated with potassium dichromate, copper sulphate and ferrous sulphate shows excellent to good fastness properties. Wool yarns dyed with turmeric (Mathur & Gupta, 2003) when subject to different concentration of natural mordant and chromium under identical mordanting conditions, shows similar colour fastness.

Application of tulsi leave extract on textiles with or without using metallic salts produces pale to dark green and cream to brown shades with adequate fastness (Patel et al, 2002). Silk (Maulik & Pal, 2005) fabric being mordanted with magnesium sulphate produces lower depth of shade, whereas copper sulphate produces highest depth. It is reported (Bhattacharya & Shah, 2000) that the colour depth of dyeing textiles can be improved by using different metal salt as mordants. Pre-mordanting and post- mordanting (Das et al, 2005 & 2006) employing ferrous sulphate and aluminium sulphate improve the colour uptake, light fastness and colour retention on repeated washing for application of many natural dyes on textiles. The use of such mordants, however, does not improve wash fastness property of textile substrate dyed with pomegranate. Dyeing of wool (Chan et al, 2000) with four varieties of tea employing different mordant shows that coloured protein fibres became blackish, when ferrous sulphate is employed as mordanting agent. The effect of mordants on yellow dyes such as kapila, onion, tesu, and dolu are also reported.

Tin, as mordant imparts good wash fastness to cotton dyed with golden rods; chrome for mariegold dyeing and alum and tin for dyeing with onion skins (Vastard et al, 1999). Turmeric dye (Devi et al, 1999) can be applied for dyeing cotton fabric by using different mordants like tannic acid, alum, ferrous sulphate, stannous chloride and potassium dichromate to obtain various shades of colour. The use of gluconic acid as a ligand for complexing iron (II) salts and for vat dyeing of cotton has been studied. It is reported (Chavan & Chakraborty, 2001) for use of iron (II) salts complexed with ligands as tartaric acid and citric acid for the reduction of indigo at room temperature and subsequent cotton dyeing. Wash fastness (Kumar & Bharti, 1998) and light fastness (Sudhakar et al, 2006) can be increased by the use of metal salts or tannic acid on cotton fabrics. Cotton yarns treated with acalypha (Mahale, 2002) dye after pre-mordanted with potash alum, potassium dichromate, copper sulphate and ferrous sulphate shows excellent fastness rating.

Pre-mordanting route favours dyeing of jute (Samanta et al, 2003) fabric with direct type of natural dyes, when aluminium sulphate is used as a mordant, while simultaneous mordanting route gives better results for madder on cotton with the same mordant. It has also been proposed that alum (Potsch 1999) and aluminium sulphate should be used as mordants in dyeing with natural dyes, as their environmental toxicity is almost nil. The effects of different natural and chemical mordants like aluminium sulphate, tartaric acid and cetrimide on colour yield for bleached jute fabric are studied and reported (Samanta et al, 2006 & 2007). As the mordant concentration and dye concentration is increased, there is improvement in the light fastness by ½ to 1 grades. Different type of mordant and method of mordanting significantly affected the rate and extent of photofading. The use of copper or ferrous sulphate give high resistance to fading, whereas stannous chloride or alum did not. On the other hand, light fastness is improved when post-mordanting is conducted with copper or ferrous ion, but pre-mordanting is superior in the case of stannous chloride or alum as investigated and reported (Gupta et al, 2004). Harda-tartaric acid combination is found to be the best followed by tannic acid-harda and tartaric acid–tannic acid combinations. Synergistic effect of mordant is observed while using the binary combinations of mordants. Meta-mordanting gives the best results for harda-tartaric acid and tartaric acid-tannic acid combinations, while pre-mordanting gives the best results for tartaric acid-harda combination as studied (Deo & Paul, 2000a & 2000b). The colour fastness properties of goldendrop root dyed on wool (Bains et al, 2005) are studied using combinations of mordants such as alum: chrome, alum: copper sulphate, alum: ferrous sulphate, chrome: copper sulphate, chrome: ferrous sulphate and copper sulphate : ferrous sulphate in ratio of 1:3, 1:1 and 3:1. Studies are available for the effects of combination of mordant on colour fastness properties of cotton dyed with peach (Bains et al, 2003). There are lots of literatures available for mordanting prior to normal dyeing and the effects of mordants on colour fastness properties, shade development and other physical properties when applied singly (Fatima & Paul, 2005; Deo & Paul, 2003) or in combination (Yu et al, 2005) on cellulosic, protenic and synthetic fibres. An effective double pre-mordanting system is recommended (Samanta et al, 2006, 2007, 2010, 2011) for dyeing jute and cotton fabric using harda (as mordant assistant cum catcher) and aluminium sulphate (metallic mordant) without intermediate drying after mordanting, facilitating wet on wet dyeing in jigger. The myrobolan (harda) powder is soaked in water (1:10 volume) for overnight (12h) at room temperature to obtain the swelled myrobolan gel. This gel is then mixed with a known volume of water and heated at 80ºC for 30 min. The solution is then cooled and filtered in a

60 mesh nylon cloth and the filtrate is used as final harda solution (10-40%) for 1st mordanting, using MLR of 1:20 (for dyed in bath) or 1:5 (for jigger). Pre-wetted conventional H2O2 bleached jute and cotton fabrics are separately treated with the harda solution in separate bath initially at 40-50ºC and then the temperature is raised to 80ºC. The mordanting is continued for 30 min. After the harda mordanting, fabric samples may be subjected to immediate wet on wet dyeing or may be dried in air without washing for storing purpose. Second mordanting is carried out using 10-40% of any one of the chemical mordants, (e.g., aluminium sulphate, potash alum, ferrous sulphate, stannous chloride and EDTA) at 80°C for 30 min using ML ratio of 1:20 (for dyed in bath) or 1:5 (for jigger). After the mordanting, the fabric samples may be finally dried in air without washing

for storing purpose to make them ready for subsequent natural dyeing or may be subjected to immediate wet on wet dyeing without drying.

PRINCIPLE OF NATURAL DYEING

Most of the natural dyes have no substantivity on cellulose or other textile fibres without the use of a mordant. The majority of natural dyes need a mordanting chemical (preferably metal salt or suitably coordinating complex forming agents) to create an affinity between the fibre and dye or the pigment molecules of natural colourant. These metallic salts as mordant form metal complexes with the fibres and the dyes. After mordanting, the metal salts anchoring to the fibres, attracts the dye/organic pigment molecules to be anchored to the fibres and finally creates the bridging link between the dye molecules and the fibre by forming coordinating complexes. Aluminium sulphate or other metallic mordants anchored to any fibre, chemically combine with certain mordantable functional groups present in the natural dyes and bound by coordinated/covelent bonds or hydrogen bonds and other interactional forces as shown below:

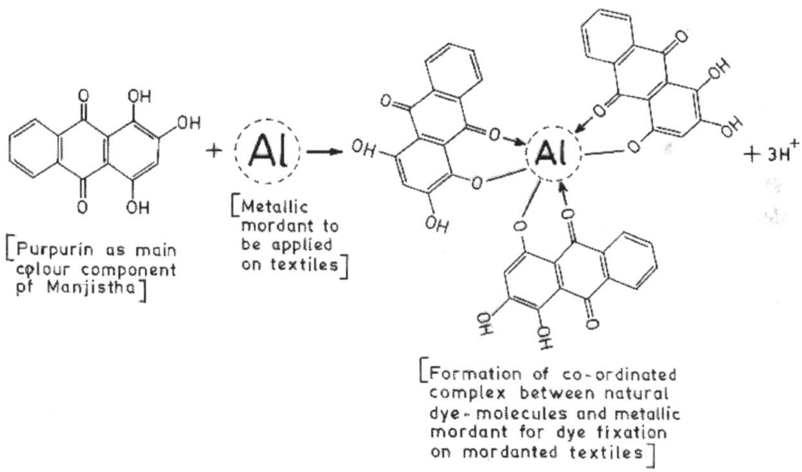

Mechanism of fixation of natural dyes through mordants

Thus, for proper fixation of natural dyes on any textile fibre, mordanting is essential in most of the cases. The said mordanting can be accomplished either before dyeing (pre- mordanting), or during dyeing (simultaneous mordanting) or after dyeing (post- mordanting).

Conventional methods of natural dyeing

Dyeing can be carried out in an alkaline bath, acidic bath or in a neutral bath. There are various reports available on different methods of mordanting on different fibers such as cellulosic, protenic and synthetic for dyeing with different natural dyes. Dyeing of cotton and silk with babool, tesu, manjistha, heena, indigo, mariegold etc is reported (Gulrajani et al, 1992; Saxena et al, 2001; Vankar et al, 2001; Nanda et al 2001; Patel & Agarwal, 2001). Various kinds of shades like black to brown, green to yellow to orange, etc can be obtained by application of different mordants.

Preparation of cotton fabric and dyeing with natural dyes

Cotton is purely cellulosic fibre and found throughout the world with many varieties and qualities. In general, cotton fibre based textiles is desized (for woven fabric only), scoured and bleached as preparatory process before dyeing with synthetic dyes. In many places of world, the age-old process followed in preparing a cotton cloth and its dyeing with natural dyes followed by artisan/cottage level dyers is given below (Mohanty et al, 1987):-

a. Dunging - The cloth is soaked for one night in a solution of water and fresh dung.

b. Washing - Next morning, cloth is thoroughly washed, rinsed and water sprinkling is continued over the cloth at short interval until evening, then it is finally washed and dried

c. Steaming– Then the cloth is steamed for one night in an ordinary Khumb or washerman's steaming pot

d. Steeping in alkaline lye- The cloth is soaked in a mixture of water, oil [castor oil or gingili oil], and alkali (sodium carbonate or soda known as sajikar or papadkhar).

e. Rinsing- Cloth is then again rinsed thoroughly and spread out to dry.

f. The last two processes are repeated for several days, the details varied in different localities, but generally from 3 to 7 days. In specific case, the cloth is kept in the solution for sometimes, and then taken out, rinsed and dried twice daily.

g. Washing- The cloth is then finally washed in clean water, but not so thoroughly as to remove the whole of the oil, and finally dried in air under the sun.

h. Galling-The cloth is then soaked in a solution of harda (haritaki) or myrobolan (Terminala chebula) extracts. Behda or bahedas (terminala belerica) is also used instead of harda. The period during which the cloth is kept in the harda extract varied in different places but it is continued until the fabric assume a yellowish tint.

i. Drying- The cloth is spread or wrung out for drying.

j. Mordanting- The cloth is then pre-mordanted by dipping it in a solution of potash alum and water. In some places, gum or a paste of tamarind seed (tamarind kernel powder) is added to make it sticky. In some parts of kutch, fuller's earth is also used by some dyers. The cloth is thus ready for subsequent dyeing.

k. Dyeing- For dyeing the cloth is generally boiled with an aqueous extracted solution of the natural dye until all the colouring matter is absorbed by the cloth.

l. Further dunging- In some places, the cloth is further soaked in dung for one night and batched before final wash and dry.

m. Drying- The dyed fabric is next washed and spread out to dry gradually in air under the sun. Water is sprinkled at certain interval over the cloth, so as to brighten the colour, this process is continued for 2-4 days.

n. Finishing- If required, the cloth is finally starched by dipping it in a paste of rice or wheat flour, or in a solution of babool gum and then dried.

However, now a days, many small scale dyers/export oriented units follow much shorter economical and standard recipe based optimized processes for natural dyeing of cotton yarns/fabrics. Before natural dyeing usual method of desizing (acid bath), scouring (soap & soda) and H2O2 bleaching are followed. Well prepared cotton textiles are then mordanted (single or double mordanting using harda and aluminium sulphate individually or in combination) before subjecting to dyeing with aqueous extract of selective natural dyes at standardized condition of process variables of dyeing. For e.g., the dyeing conditions may be as follows : dyeing time, 30 -120 minutes (depending on shades); dyeing temperature, 70- 100°C; material to liquor ratio, 1:20 -1:30; concentration of natural dye, 10-50% (owm) or more; common salt concentration, 5-20g/L and pH, 10-12. In each case after the dyeing is over, the dyed samples are repeatedly washed with hot and cold water and then finally, the dyed samples are subjected to soaping with 2g/L soap solution at 60 ?C for 15 min, followed by repeated water wash and line dried. For improving its wash fastness, treatment with eco- friendly cationic dye fixing agent is advisable.

Dyeing process for natural colouration of wool and silk fibre

Wool and silk are natural protein fibres and are available in wide variety having varied qualities. Both the fibres has complex chemical structure and very much susceptible to alkali attack (at pH >9). Hence, dyeing of these fibres with natural colours needs special care to avoid fibre damage by alkaline pH. Moreover, both wool and silk contain both amino and carboxylic functional groups. While, unlike silk, wool contains equal number of amino and carboxylic groups held together as salt linkages which bridge the main peptide chains. Therefore, in aqueous solution, wool carries no net charge. However, silk fibre has a slightly cationic character with the isoelectric point at about pH 5.0. Also, unlike wool fibre, silk is less sensitive to temperature. Therefore, selection of mordants, conditions of mordanting, pH and other conditions for dyeing, necessary cares are to be taken for colouration of these textile fibres for mordanting and application of natural dyes.

Wool and silk fibre based textiles can be dyed with different natural colours mostly through pre-mordanting or post mordanting system. Mordanting is done with tannin rich natural source chemicals like harda, gall nut etc and/or metal salts like, alum, aluminium sulphate, ferrous sulphate etc. Depending on shade depth requirement, mordant and dye concentration are to be determined. Dyeing conditions for a particular fibre-mordant–dye system need to be optimized by study of dyeing process variables before bulk dyeing.

In pre-mordanting system, these animal fibre based textiles are selectively mordanted (single or sequential double mordanted) with 5-20% (owm) mordant at 80-90°C for about 30- 40 minutes having ML ratio 1:5-1:20 and can be taken for subsequent dyeing generally without washing. The pre-mordanted samples are entered in the dye bath (generally acidic dye bath) at 50-60°C and ML ratio 1:20, raise temperature upto the optimum dyeing temperature (may be 90°C) and dyeing is then continued for further 30 to 40 minutes followed by thorough rinsing, soaping and washing. However, in post mordanting system of applying natural colour on wool and silk textiles, dyeing is done at optimum dyeing condition and the dyed samples are dip into a mordant bath containing mainly 1-2% metal salt (owm) for true colour development followed by rinsing, soaping and thorough washing.

9.1.1 Dyeing process for natural colouration of jute and other lignocellulosic fibres Application of natural colours on ligno-cellulosic fibres through double pre-mordanting system is found to be the best method, e.g., 10-20% harda treatment followed by 10-20% alum treatment or aluminium sulphate treatment is best suited mordanting system for subsequent dyeing of lingo-cellulosic fibre based textiles with different natural colours/dyes. Conventionally bleached and double pre-mordanted jute and linen textiles without wash, is dyed with tesu, madder, catechu, pomegranate rind, babool, jackfruit wood, haldi, marie gold, red sandal wood etc individually and in mixtures. Most of the above said natural dyes are applied in alkaline pH like 11-12 at higher temperature like 80- 90°C having ML ratio 1:20 for 60 to 90 minutes. After dyeing, the dyed samples are washed and soaping is carried out at 60°C for 15 minutes. For getting higher fastness to wash and light, the samples can be further treated with 2% cationic dye fixing agents and/or 1% benztriazole.

Dyeing process for natural colouration of synthetic fibres

Different synthetic fibres like nylon, polyester etc are dyed with various source natural dyes/colourants like onion skin extract, babool bark extract, henna etc. through exhaust, HT-HP and padding methods (cold-pad-batch) with or without mordanting. Mordanting facilitates to get wide range of shades from the same source of natural dye as per requirement by variation of mordant chemicals as well as mordanting techniques. In case of dyeing synthetic fibres, dyeing is carried out at acidic pH and HT-HP dyeing technique results an overall best dyeing performance in terms of colour strength and fastness properties provided the natural colourants are stable at that high temperature employed for dyeing. Also energy consumption is higher in case of HT-HP dyeing method than exhaust method as well as cold-pad-batch method of dyeing.

Reports are available (Lokhande et al, 1998; Lokhande and Dorugade, 1999) for nylon is dyed with three different natural dyes using various mordants by two different techniques (open bath and high temperature high pressure dyeing methods), of which HT-HP dyeing is found better as compared to open bath. Application of babool bark extract on nylon substrate by cold-pad-batch and pad-dry-steam technique of dyeing can be considered as an effective eco- option and can be commercialized. HT-HP method is used for dyeing polyester fibre with pomegranate rind, catechu, nova red and turmeric (Bhattacharya and Lohiya, 2002).

9.2 Non-conventional dyeing methods

Nanotechnology is increasingly attracting worldwide attention because it is widely perceived as offering huge potential in a wide range of end uses. The unique and new properties of nanomaterials have attracted not only scientists and researchers but also businesses, due to their huge economical potential. One possible application is to directly employ pigment nanoparticles in textile coloration. Such an approach could be achieved if the nanoparticles can be reduced to a small enough size and the particles can be dispersed well to avoid aggregation of the nanoparticles in dye baths. Exhaust dyeing of cationized cotton with nanoscale pigment dispersion has recently been achieved and the results indicated that the dyeings obtained have better soft handle and more brilliant shade with reduced pigment requirement than those obtained with a conventional pigment dispersion (Fang et al, 2005).

Ultrasonic energized dyeing conditions for neem leaves gives better dye uptake, uniform dyeing, better light and wash fastness on cotton fabric (Senthikumar et al, 2002). Unconventional natural dyeing of cotton with sappan wood by ultrasound energy as well as new methods using microwave and sonicator for application of natural dye from alkanet root bark on cotton and dyeing of cotton fabrics with tulsi leaves extract by use of ultrasonic energy dyeing are also reported (Ghorpade et al, 200; Tiwari et al, 2000a & 2000b).

The application of supercritical carbon dioxide (scCO2) in the textile industry has recently become an alternative technology for developing a more environmentally friendly coloration process. scCO2 coloration technology has the potential to overcome several environmental and technical issues in many commercial textile applications such as yarn preparation, coloration and finishing. scCO2 represent a potentially unique media for either transporting chemical into or out of a polymeric substrate, because of their thermo-physical and transport properties. Supercritical fluids exhibit gas-like viscosities and diffusivities and liquid-like densities. Additionally, carbon dioxide is nontoxic, non-flammable, environmentally friendly, and chemically inert under many conditions; however, its production is remained to be cost-effective. The dissolving power of scCO2 for disperse dyes and its use as the transport media for coloration polyester is studied from all theoretical aspects at DTNW in Krefeld, Germany. Studies have revealed that the presence of intramolecular hydrogen bonds and/or the hydrophobicity of dye molecule are positive factors for better solubility in supercritical carbon dioxide, as indicated from its improved dye-uptake (Shakra et al, 1999 & 2000)

The use of scCO2 as a fluid medium for coloration of textile fibers, especially polyester, has been examined. This technology has become so promising that it has provided new opportunities to develop suitable dyes for this medium. The coloration is conducted in a stainless steel high pressure apparatus. Process and equipment are developed for textile dyeing in supercritical carbon dioxide (Kraan, 2007). A technical-scale, 100-

L dyeing machine is designed and built for polyester beam dyeing in $scCO_2$ at 300 bar.

PHYSICO-CHEMICAL STUDIES ON DYEING PROCESS VARIABLES AND DYEING KINETICS

The natural dyes have a variable chemical composition, which is influenced by a number of factors, out of which the most important are: the vegetal part of the plant where the extract is obtained from, its place and growing conditions, harvesting period, extraction operation and application or technological process followed. Many workers (Gupta, 1999) have reported some of the most significant experimental results, laboratory trials regarding techniques and best parameters for extraction and application of natural dyes, i.e. observations on varying extraction parameters, such as: extraction temperature, extraction time, extraction solvent-vegetal material ratio, type of solvent and observations on varying mordanting and dyeing parameters. However, most of such studies are concentrated on wool and silk.

The effects of dye-extraction medium, optimum concentrations of dye source material, extraction time, dyeing time, mordant concentration and methods of mordanting on silk dyed with natural dyes has been reported (Grover et al, 2005; Dixit & Jahan, 2005; Dumitrescu et al, 2005; Srivastava et al, 2006; Das et al, 2005; Agarwal et al, 1992; Bansal & Sood, 2001; Sati et al, 2003; Rose et al, 2005; Maulik & Bhowmik, 2006; Siddiqui et al, 2006). The acidic media exhibited maximum percent absorption for jatropa, lantana, hamelia and euphorbia dye, while kilmora and walnut showed good results in alkaline medium. The result obtained from different experiments lead to the optimization of a standard recipe for particular dye-mordant-fibre combination. The optimum concentration of beet sugar (Mathur et al, 2003) colourant for dyeing wool is found to be 0.03g per g of wool at pH 4.5 and temperature 97.5°C. Dyeing of wool under the optimum (Mathur et al, 2001) condition pH 4.5, colourant concentration-0.05g per gram of wool; time-60 min and treatment temperature- 97.5°C shows very good light and wash fastness properties without deteriorating the quality of wool. Optimum dyeing technique for colouration of wool by determining the optimum wavelength, dye material concentration, extraction time, dyeing time, pH, concentration of mordant etc. helps to standardize the dyeing process (Srivastava et al, 2006). Colouring component of tea shows highest affinity for both wool and silk at pH 2 to 4 in presence and absence of ferrous sulphate and aluminium sulphate as mordants (Das et al, 2005). Optimisation of dyeing process variables for wool with natural dyes obtained from turmeric has been studied and reported (Agarwal et al, 1992). The optimum conditions for development of vegetable dye on cotton from eupatorium leaves are studied and reported (Bansal & Sood, 2001). The optimization of wool by using rhododendron arboretum as a natural dye source is reported (Sati et al, 2003). The effect of process variables on colour yield and colour fastness properties for application of selective natural dyes for different textiles are also studied and reported (Rose et al, 2005; Maulik & Bhowmik, 2006; Siddiqui et al, 2006).

The dyeing absorption isotherm for wool, human hair, silk, nylon and polyester is found to be linear indicating a partition mechanism of dyeing for application of juglone as natural dye (Gupta & Gulrajani, 1993). The slope of isotherms increases with the increasing temperature in all cases. ΔH and ΔS values are positive for all the dyeings. The apparent diffusion coefficient is highest for wool and lowest for silk. For investigation of the conditions of extraction and application of african mariegold on silk yarn, optimum conditions are found to be 60 minutes dye extraction time, 30 minutes mordanting, 30 minutes dyeing using mixtures of 5% potash alum, 1% potassium dichromate and 1% copper sulphate as mordants (Mahale et al, 1999). Studies of dyeing absorption isotherm, heat of dyeing, free energy and entropy of dyeing for red sandal wood (Samanta et al, 2006) and jackfruit wood (Samanta & Agarwal, 2008) reveals that both the dyeing process follows a linear Nernst absorption isotherm. An adsorption and thermodynamic study of lac dyeing on cotton pre-treated with chitosan showed Langmuir isotherm (Rattanaphani et al, 2007). The colour can vary significantly for indigo dyeing on denim yarns as the result of the variation of parameters (Kin et al, 2007).

COMPATIBILITY OF MIXTURE OF SELECTIVE NATURAL DYES

Newer shades can be achieved by applying mixture of natural dyes. For the use of mixture of natural dyes, the dyers must know whether the natural dyes are compatible with each other or not. For test of compatibility of a pair of natural dyes, bleached and pre-mordanted samples are dyed in two different sets of progressive depth of shade of binary mixture of natural dyes as follows:-

Bleached and pre-mordanted textiles are dyed with the specific binary mixtures of natural dye pair. The dyeing is started at 40°C and the temperature is gradually raised to 100°C taking total time of 60 min (approximately) to raise this temperature at a heating rate of around 1°C/min, using precision temperature controller open bath Laboratory beaker- dyeing machine.

In set I, (progressive depth of shade developed by varying dyeing time and temperature profile during dyeing), for 1% shade with binary mixture (50:50) of a pair of dyes, pre- mordanted samples are separately dyed for different dyeing period (10, 20, 30, 40, 50 and 60 min), by withdrawing separately dyed samples from the dye bath at the intervals of 10 min from 50?C onwards maintaining the heating rate of around 1°C/min. The penultimate sample is taken out after 50-60 min at 90?C and the last one at the end of the dyeing carried out for 60 min at 100°C.

In set II (progressive depth of shade developed by varying total concentrations of dye mixture using 20-100 parts of 1% shade depth using purified natural dye-stuff colourants) for a pair of binary mixture dyes, pre-mordanted samples are separately dyed at increment of twenty percentage points by applying 20 -100 parts of 1% dye (on weight of fabric) for each pair of natural dye-mixture taken in equal proportion (50:50) at 100?C for 60 min. For both Set I and Set II, after dyeing, all the dyed samples are subjected to normal washing, soaping, and rinsing before final air-drying. The corresponding surface colour strength (K/S value) and the differences in the CIELAB coordinates namely, ΔL, Δa, Δb and ΔC for all dyed fabric samples for Set I and II obtained indicate for the samples dyed with using purified natural dyes, the colour yield (K/S), lightness/darkness (ΔL), redness/greenness (Δa), yellowness/blueness (Δb) and differences in saturation/chroma (ΔC) values with respect to the standard un-dyed sample, which are measured and obtained from separate measurement of the same using the reflectance spectrophotometer and associated software and computer. Plots of K/S vs ΔL and/or ΔC vs ΔL, i.e., two sets of curves obtained for the said two sets (Set-I & Set-II) of dyed samples indicate the nature of compatibility by closeness of the pattern of the two sets of curves (Konar, 2011). However, another easy method of Relative Compatibility Rating (RCR) method (Samanta et al, 2009) has been established based on differences between lowest and highest colour difference index (CDI) values [CDI = (ΔE X ΔH) / (ΔC X MI)] for a binary mixture of natural dyes dyed in different proportions with fixed shade % at standardized dyeing conditions.

COLOUR FASTNESS PROPERTIES OF NATURAL DYES

Colour fastness is the resistance of a material to change in any of its colour characteristics or extent of transfer of its colourants to adjacent white materials in touch or both for different environmental and use conditions or treatments like washing, dry cleaning etc or exposure to different agency heat, light etc. Fading means changes in the colour with or without loss of depth of shade for exposure to particular environment/agency/treatments either by lightening or darkening of the shades. Bleeding is the transfer of colour to a secondary material in contact accompanying white fibre material of similar/dissimilar nature. The colour fastness is usually rated either by loss of depth of colour/ colour change in original sample or it is often expressed by staining scale meaning that the accompanying material gets tinted/stained by the colour of the original fabric, when the accompanying white fabrics of similar/dissimilar nature are either in touch/ made to touch by some means of test procedure/protocol.

LIGHT FASTNESS

An extensive work has been carried out to improve the light fastness properties of natural dyed textiles. A comprehensive review on different attempts taken for improving colour fastness properties of dyes on different textile fibres by different means is reported (Cook, 1982). The said review includes tannin-related after-treatments for improving the wash fastness and light fastness of mordantable dyes on cotton; some of these treatments might be applicable to selective/specific natural dyes.

Most of the natural dyes have poor light stability (as compared to that of the best synthetic dyes), and hence the colours in museum textile are often different from their original colours. The relative light stability of a range of dyes has been reviewed (Padfield & Landi, 1966) along with studies involving change in qualitative fashion. These changes in colour are studied quantitatively (Duff et al, 1977) where it is expressed the changes in terms of the Munsell scale and also in CIE colour parameters. Wool dyed with nine natural dyes is exposed in Microscal MBTF fading lamp. The fastness ratings are similar to those found earlier in day light fading. After rating by the blue wool standards for light fastness rating, yellow dyes (old fustic, persian berries) shows poor light fastness between 1-2; red colours like cochineal(tin mordant), alizarin (alum and tin mordant), lac (tin mordant) shows better light fastness between 3-4 ; indigo shows light fastness 3-4 or 5-6 (depending on the mordant) ; and logwood black (chrome mordant) shows light fastness 4-5 or 6-7 (other mordants). It is also reported the effects of chemical structure of natural dyes on light fastness and other colour fastness properties (Gupta, 1999a & 1999b).

A large proportion of natural dyes are, of course, mordant dyes. There is strong influence on nature, type and concentrations of mordants on wash and light fastness grades. The influences of different mordants are found to play important role in fading of 18 yellow natural dyes (Crews, 1982). Where wool dyed with different natural dyes specimens are exposed to a xenon arc lamp for assessing their light fastness upto 8 AATCC Fading Units equivalent to BS-8B Blue wool standards. The corresponding colors changes after exposure to xenon arc lamp are also assessed in each case.

Turmeric, fustic and marigold dyes faded significantly more than any of the other yellow dyes. However, use of tin and alum mordants resulted in significantly more fading than the same for use of chrome, iron, or copper mordant. Thus the type of mordant is found to be is more important than the dye itself in determining the light fastness of natural colored textiles.

Natural dyes that may have been used in the traditional Scottish textile industry have been described (Grierson, 1984) that the light fastness of such dyes on wool has been compared with those of dyeing with 'imported' dyes to similar shades (Duff et al, 1985; Grierson et al,1985), again using Microscal fading MBTF lamp.

Numerous attempts (Samanta et al, 2006, 2010 & 2011; Hofenk, 1983; Oda, 2001; Cristea & Vilarem, 2006; Lee, 2001; Micheal & Zaher, 2005; Gupta, 2001) has been made to improve the light fastness of different textiles fabric dyed with natural dyes which include the effects of various additives on the photofading of carthamin in cellulose acetate film, critical examination of fading process of natural dyes with a view to determining the original colours of faded textile etc. The rate of photofading is remarkably suppressed in the presence of nickel hydroxy-arylsulphonates, while the addition of UV absorbers afforded little retardation in the rate of fading.

WASH FASTNESS

With a view to examine and improving wash fastness (Duff et al, 1977), tests are carried out under standard condition (50°C) and also at 20°C with a washing formulation used in conservation work for restoration of old textiles. Some dyes undergo marked changes in hue on washing, shown to be attributed to even small amounts of alkali in washing mixtures, high-lighting the necessity of knowing the pH of alkaline solutions used for cleaning of textiles dyed with natural dyes. As a general rule, natural dyes (on wool) have only moderate wash fastness as assessed by the ISO 2 test. However, logwood and indigo dyes exibit better fastness when applied to different textiles. The nature of detergent solution suitable for conservation of natural coloured art work has been examined (Hofenk, 1983). A liquor containing 1g/l of sodium polyphosphate is found to be best resulting marginal changes in hue with natural dyes applied on wool or silk (Duff et al, 1977). The small increase in cleaning efficiency attributable to the alkali must be balanced against possible colour change in the natural dyes, apart from possible damage to the protein fibre under alkaline conditions.

In the ISO 2 test, the fastness of the indigo and logwood is superior to that of the native natural dyeing such as privet berries and water lily root respectively, but in the comparison of native and imported yellow, reds, red/purples, greens and browns, there is little difference between the two groups (Duff et al, 1977). It is found from a recent report that treatment with 2%CTAB or sandofix-HCF improve the wash fastness to nearly 1 unit and treatment with 1% benztriazole improved the lightfastness of dyed jute textiles nearly half to one unit (Samanta et al, 2006; 2007; 2010 & 2011; Samanta & Agarwal, 2008)

RUBFASTNESS

Rub fastness of most of the natural dyes have been found to be moderate to good and dose not require any after treatment. Jackfruit wood, manjistha, red sandal wood, babool, mariegold etc have good rubfastness (Samanta et al, 2006; 2007; 2010 & 2011). Good rub fastness is seen for mariegold on cotton, silk and wool (Sarkar et al, 2005; Sarkar, 2006). Good rub fastness (dry and wet rubfastness) is reported for silk dyed with acalypha and other natural dyes (Mahale et al, 1999; 2002 & 2003). Cutch and ratanjot shows moderate to good dry rub fastness but the wet rub fastness is found to be average (Khan et al, 2003 & 2006).

CONCLUDING REMARK

Most of the natural dyes/ colour are eco-safe, except a few. Some of the natural colours are not only eco-safe, but also has added value for its medicinal effects on skin and are more than skin friendly. Textile dyers must know the chemistry of these natural colours and its added advantages of medicinal; values. Use of suitable binary opr ternary mixtures of similar or compatible natural dyes for colouring natural eco-friendly textiles in variety of soothing / uncommon shades with eco-friendly mordants and finishing agents are the most desireable product of the customers for future. The non-reproducibility and poor colour fastness etc, have been partly solved by many researchers' continuous effeorts in this endevour. So, a textile dyer must know the effects of variability for extraction, mordanting and dyeing and should follow only the standardized recipe for selection fibre-mordant- natural dye system to get reproducible colour yield and colour matching besides to follow different eco-friendly ways to improve colour fastness to a possible extent.

This chapter is a highlight of all these efforts towards popularizing natural dyeing not only in small/cottage dyers but also textile yarn/fabrics industrially using common dyeing machine. Some of the study mentioned above explained this clearly. So, natural dyeing/colouration of textiles by industrial processes in large scale dyeing unit is now a reality in the textile market of eco- friendly textiles.

REFERENCES

1. Agarwal A, Goel A & Gupta K C, *Textile Dyers and Printer*, 25(10), (1992), 28 Agarwal A, Garg A & Gupta K C, *Colourage*, 39(10) (1992) 43.
2. Agarwal A, Paul S and Gupta K C , *Indian Text J*, (1) (1993) 110. Bansal S & Sood A, Textile Magazine, 42 (8), (2001), 81
3. Bhattacharya S K, Dutta C & Chatterjee S M, *Man-made Textiles in India*, (8) (2002) 207.
4. Blanc R, Espejo T, Montes A L, Toress D, Crovetto G, Navalon A & Vilchez J L, *J Chromatography A*, 1122 (2006) 105
5. Balankina G G, Vasiliev V G, Karpova E V & V I Mamatyuk, *Dyes Pigm*, 71(2006) 54. Bhattacharya N, in Proceeding of convention of Natural Dyes edited by Deepti Gupta &
6. M.L Gulrajani, Department of Textile Technology, IIT Delhi, 1999, 134 Benencia F & Courreges M C, *Phytomed*, 6 (2) (1999) 119.
7. Bhattacharya S K, Chatterjee S M & Dutta C , *Man-made textiles in India*, (2004) 85 Bain S , Singh O P & Kang K , *Man-made textiles in India*, 45(8) (2002) 315.
8. Bhattacharya S D & Shah A K , *J Soc Dyers Color*, 116 (1) (2000) 10. Bains S , Kaur K & Kang S , *Colourage*, 52 (5) (2005) 51.
9. Bains S, Kang S & Kaur K , *Man-made Textiles in India,* 46(6) (2003) 230. Bhuyan R, Saikai C N &Das K K, *Indian J Fibre Text Res*, 29(12) (2004)429, 470.

Chapter 8

A Facile Energy and Water-Conserving Process for Cotton Dyeing

Rachana S. Harane • Neha R. Mehra •Priti B. Tayade • Ravindra V. Adivarekar

Department of Fibres and Textile Processing Technology,Institute of Chemical Technology (ICT), Matunga,Mumbai 400019, Maharashtra, India

ABSTRACT

Nowadays, all research efforts in the field of wet processing of textiles are directed towards shortening and simplification of the processing sequence. To minimise water and energy consumption, it has become necessary to combine several textile processing stages to reduce the number of operations or shorten the process time. An attempt was made to combine the pre-treatment and dyeing processes with optimal quantities of textile auxiliaries in single bath. In this process, individual stages of pre-treatment like desizing, scouring and bleaching followed by dyeing using direct dyes were carried out in a single bath without the usual intermediate washings. This process was carried out without replacing the liquor bath until the end of dyeing. An attempt was also made to dye a woven fabric at different stages, viz., grey, desizing and scouring. These efforts were made to conserve water and energy required for that particular pre-treatment process. The depth of dyeing and the colour fastness properties of all these dyed fabrics were then compared with a bleached fabric which was dyed with conventional method. The properties were at par with the conventional method. Moreover, the combined process saved 83.1 % water, 88.6 % energy (thermal) and 79.3 % time than the conventional processes. Thus, the process has been shown to be ecological, economical and energy conserving.

Keywords: Pre-treatment Dyeing Direct dyes Single bath Depth of dyeing Colour fastness Energy conserving

INTRODUCTION

The conventional three-stage desizing, scouring and bleaching operations are exceptionally intensive in water, energy and chemicals which ultimately influence the process cost. At the same time, limited availability of water is also a challenge. The water requirement varies from mill to mill depending on the process sequence they follow. The quantity of water required for textile processing depends on the textile process, fabric produced, the equipment type and dyestuffs.

Conventional preparation of cotton-based textiles usually involves desizing of the starch (size) applied to the yarn during weaving, scouring of the material to remove the natural hydrophobic impurities such as fats and waxes, followed by a bleaching process to impart whiteness by destroying the natural colouring matter [1–3]. The water usage required for different wet-processing stages in a typical cotton textile mill is given in Table 1[4].

Table 1. Water required for cotton textile wet processing [4]

Processes	Water requirements in L/1,000 kg of products
Desizing	2,500–21,000
Scouring	20,000–45,000
Bleaching	2,500–25,000
Dyeing	10,000–300,000
Printing	8,000–16,000

From Table 1, it is clear that the pre-treatment process, i.e. desizing, scouring and bleaching, involves more water usage. Also, about 55 to 60 % energy consumed in the textile industry is used in various pre-treatment stages [5]. The high consumption of water and energy for textile wet-processing sector emphasises the urge for improvement of process, i.e. to adopt cost effective practices to reduce the use of scarce resources like water and energy. Hence, the logical approach to combine desizing, scouring and bleaching operations to save energy and water was put forward by many researchers [6–10].

Various researchers have also worked on combining pre-treatment and dyeing processes using various enzymes. Quardt and Kuhl [11] have attempted to use enzymes from the beginning to the end of the finishing processes of textile substrates. Hardt et al. [12] carried out single-bath bleaching and dyeing processes with ecologically optimal textile auxiliaries. These are processes in which individual stages of treatment, viz., acid demineralisation, alkaline scouring, hydrogen peroxide bleaching and the elimination of residual peroxide with subsequent reactive dyeing occur without the normal intermediate washings. Losonczi et al. [13] studied the conventional and bio-preparation scouring processes of cotton fabrics and also studied the bleaching and dyeing with reactive dyes of such pre-treated fabrics. Kokol and Golob [14] performed biopreparation of raw cotton using enzymes and also carried out reactive dyeing in the same bath. Oner and Sahinbaskan [15] carried out all of the stages viz., desizing, scouring, bleaching and dyeing in a single bath without replacing the process water until the end of the dyeing. They named the process 'Rapid Enzymatic Single-bath Treatment' (REST), since it was completed almost in half of the conventional dyeing time.

In this work, we take further steps to challenge the conventional preparatory and the dyeing processes. The aim of the work was to ensure efficient reduction in water consumption in the cotton wet-processing sector. There is a tendency to use more amount of water than required while removing the residual chemicals during rinsing [16]. To fulfil the objective, an unprocessed grey fabric was subjected to combined pre-treatment and dyeing process which was carried out using all the conventional chemicals in the same bath. Successful attempt were also made to dye fabrics at grey stage, desized stage and scoured stage. This was carried out to check the dyeing efficiency of a grey fabric after various steps of pre-treatment. Through this attempt, we tried to question the practical need of pre-treatment processes given to a grey fabric. These processes impart hydrophilicity to the otherwise hydrophobic grey fabric. We were able to dye a grey unprocessed fabric with acceptable and almost comparable properties as of a RFD fabric in our work. Though the absorbency time of grey dyed fabric is more, one can utilise these fabrics depending on their end use, e.g. upholstery fabric, technical textiles, etc. where a good absorbency is not of utmost importance. We also should take into notice that this can tremendously save water, chemicals and energy required for processing a fabric for making it ready for dyeing.

Thus, efforts were made towards not only reducing the water intake and minimise the intermediate washings but also to minimise the intake of chemicals wherever possible to make existing process, 'more sustainable' and 'environment friendly'.

MATERIALS AND METHODS

Materials

A 100 % grey cotton fabric used for experiments was procured from Tata mills, Mumbai, with a plain weave having a yarn count of 40 (The count of the cotton yarn is equal to the number of 840-yard skeins of cotton required to weigh 1 pound) and with a weight of 85 g/m^2. The size content on the grey fabric was 12 %.

The chemicals used were hydrochloric acid (HCl), wetting agent (Turkey Red Oil), sodium hydroxide (NaOH), sodium carbonate (Na_2CO_3), hydrogen peroxide (H_2O_2), sodium silicate (peroxide stabiliser), Glauber's salt and glacial acetic acid. All chemicals were supplied by S.D. Fine chemicals and were of AR grade. Saradye FN (dye fixing agent) was procured from Sarex Chemicals Ltd.

The direct dyes namely, CB Direct Red 5B, CB Direct Blue GLL, CB Direct Black FR, CB Direct Olive Green BL, CB Direct Yellow 5GLL and CB Direct Turquoise Blue BRLE 150 %, procured from Colorband India Pvt. Ltd were used for the study.

EXPERIMENTAL METHODS

Conventional methods

Desizing process

Initially weighed grey cotton woven fabric was taken and treated with 2 % HCl and 1 % wetting agent at a temperature of 50–60 °C for 2 h, excluding the time required for heating, which is 15 min at a heating rate 2 °C/min, with a material to liquor ratio (MLR) of 1:30 in a standing bath. After this, the sample was given a cold wash for 5 min followed by hot wash at 60 °C for 10 min. The fabric was given a neutralisation treatment with dilute alkali at 30 °C for 10 min and dried. Weight loss was measured. The fabric sample was also given Tegewa rating to ensure proper size removal.

Scouring process

Initially weighed desized cotton woven fabric was treated with 10 g/L NaOH, 5 g/L Na_2CO_3 and 2 g/L wetting agent at boil for 2–3 h, excluding the time required for heating, which is 35 min at a heating rate 2 °C/min, at MLR of 1:30. After this, the sample was given cold wash for 5 min followed by a hot wash at 60 °C for 10 min and dried and weight loss and absorbency was calculated.

Bleaching process

Initially weighed desized and scoured cotton woven fabric was taken and treated with 5 g/L H_2O_2, 2–5 g/L Sodium silicate and 2–5 g/L Na_2CO_3 at a temperature of 80–85 °C for 1 h, excluding the time required for heating, which is 27 min at a heating rate 5 °C/min at an MLR of 1:30. After this, the sample was given cold wash for 5 min followed by a hot wash at 60 °C for 10 min and was neutralised by acetic acid at 30 °C for 10 min. The fabric was dried and measured for Whiteness Index to ensure complete decolourisation of coloured impurities.

Direct dyeing of cotton fabric

A dye bath was prepared at room temperature by adding bleached cotton fabric and required amount of Glauber's salt. Sodium carbonate was added only for black shades to get a pH of 7–8. Then, required amount of dye was pipetted out from stock solution (1 %) into dye bath. The temperature of the dyebath was raised to 100 °C by keeping 1.5 °C/min temperature gradient and the process was carried out for 1 h in rota dyer at an MLR of 1:20. After this, the bath was drained and the fabric was rinsed thoroughly with cold water for 5 min followed by treatment with 1 % dye fixing agent at 60 °C for 30 min. Then, the fabric was rinsed and dried.

Modified method

In the modified method, pre-treatment and dyeing was carried out in a single bath. A dye bath was prepared at room temperature using 1 g/L wetting agent, 5 g/L NaOH, 3 mL/L H_2O_2, 50 % w/v (i.e. 50 g of hydrogen peroxide was weighed and was diluted up to 100 ml of water.) 1 g/L sodium silicate (peroxide stabilizer). The required amount of direct dye was also added in the same bath along with 15 g/L Glauber's salt. Sodium carbonate was added only for black shades to get a pH of 7–8. To this bath, grey woven fabric was added. Temperature was raised to 100 °C by keeping 1.5 °C/min temperature gradient and processing was carried out for 1 h in rota dyer at an MLR of 1:20. The fabric was then rinsed thoroughly with cold water for 5 min and was treated with 1 % dye fixing agent for 30 min at 60 °C. Fabric was again given cold wash for 5 min and air dried.

Testing and analysis

Testing of desized fabric

Principle

With starch sizes, the Iodine reagent (Iodine + Potassium Iodide) produces a coloured complex of intense violet blue colour. The coloured complex formed is a two-dimensional complex, where iodine, the linear chain, is included in a two-dimensional structure of the starch. This test is practical because it directly indicates either the presence of the starch by a characteristic deep blue colouration or the absence of the starch by a yellowish orange or no colouration on the fabric.

Preparation of the iodine solution

For preparation of iodine solution, 10 g of potassium iodide was dissolved in 100 mL of water and 0.65 g of iodine was added into it. This solution was stirred for complete dissolution. To the above solution, 800 mL water was added and further ethanol was added to make up the volume up to 1 L.

Application of the iodine solution

Grey fabric sample was dipped in the above-mentioned iodine solution for 1 min, was then rinsed with cold water and dabbed with filter paper. Then, this fabric was compared immediately with violet scale (Tegewa Scale) and accordingly rating was given. The scale has 1–9 rating. Rating 1 is poor, and a 9 rating is the best which indicates complete removal of size from the fabric. Commercially acceptable rating is 6–7.

Testing of scoured fabric

Absorbency test was performed using AATCC Test Method 79-1995, details of this method are given in [17].

Testing of bleached fabric

The bleached fabrics were evaluated for whiteness index using computer colour matching system (SpectraScan 5100+) supplied by Premier Colorscan Instruments Pvt. Ltd. Bleached fabric was evaluated for whiteness index using AATCC test method 110-2005 [18]. An average of two readings taken at two different positions of samples was used to calculate the indices.

Colour depth (in terms of K/S value and colour strength) and colour difference (in terms of L, a and b values)

Dyed samples were evaluated by determining K/S values as well as colour strength using computer colour matching system (SpectraScan 5100+) supplied by Premier Colorscan Instruments Pvt. Ltd. An average of two readings which was taken at two different positions of samples was used to calculate the Kubelka–Munk, i.e. K/S function which is given by,

$$K/S = (1 - R)^2/2R$$

where, R the reflectance at complete opacity

K the absorption coefficient

S the scattering coefficient

The colour difference (d*E*) can be visualised as the distance between the standard or reference and the sample points plotted in the *L a b* colour space. *L* values have a range of 0–100, 100 indicates lightness or white and 0 indicates darkness or black. Negative values of '*a*' indicate green while positive values indicate red, whereas negative values of '*b*' indicate blue and positive values indicate yellow. The *L a b* system adopted from the CIE in 1976 is the most widely used system today [19].

$$\mathrm{D}E = \left(\mathrm{D}L^2 + \mathrm{D}a^{2+}\mathrm{D}b^2\right)^{1/2}$$

where D*L* = *L* (sample) – *L* (standard)

D*a* = *a* (sample) – *a* (standard)

D*b* = *b* (sample) – *b* (standard)

From the CIELAB coordinates, one can compute *C* (chroma) and *h* (hue). The Chroma or saturation (*C*) is the distance between achromatic point and colour is calculated from '*a*' & '*b*' using following equation;

$$C = \left(a^2 + b^2\right)^{1/2}$$

The point $a = 0$ and $b = 0$ lies at the centre of the chromaticity diagram. Lower is the value of '*a*' and '*b*', higher is the achromaticity and lower is the purity. On the other hand, higher the values of '*a*' and '*b*' (ignore the negative signs), purer or more saturated or brighter is the colour.

$$h = \tan^{-1}\,(b/a) \text{ in degrees}$$

Positive values of chroma (or saturation) indicate more saturated colour (i.e. brighter) and negative values indicate less saturated colour (i.e. duller). Hue is an angle and can also be termed as purity of colour [19].

Washing fastness evaluation

Washing fastness of dyed fabrics was done using ISO 105-C03 [20]. In this method, the specimen is sandwiched between a wool fabric and a bleached fabric and is stitched. Then, it is placed in the container along with 5 g/L soap solution and 2 g/L sodium carbonate with an MLR of 1:50. It was run at a temperature of 60 ± 2 °C for 30 min. Then, the sample was removed, rinsed twice under tap water for 10 min and squeezed. Then, it was air dried and the required testing was carried out, i.e. change in the colour of the dyed specimen, staining on cotton fabric and staining of wool fabric.

Light fastness evaluation

The light fastness of the dyed samples was tested on Q-sun light fastness tester and the fastness was graded for the colour change when the samples were partially exposed to the Xenon lamp for 8 h. The method adopted is ISO 105-B02 (which is similar to AATCC 16) [21].

Rubbing fastness evaluation

Rubbing fastness of the dyed samples was determined using automatic Crockmeter. The method adopted is from ISO 105 X12 evaluation procedure [22]. Both dry rubbing and wet rubbing of the dyed fabric were assessed. Suitable testing device for determining the colour fastness to rubbing consists of a reciprocating straight line rubbing motion and a rubbing finger. The rubbing finger shall exert a downward force of 9 ± 0.2 N moving to and fro 20 times, 10 times in each direction in a straight line along a path of 10 ± 3 cm track. A bleached white fabric is attached to the rubbing finger and is rubbed against the dyed sample. While assessing wet rubbing fastness, the fabric is moistened with water and then rubbed. The staining of the cotton rubbing cloths is assessed with the grey scale for staining.

Results and discussions

Desizing was carried out using conventional method as described in the experimental methods, weight loss and Tegewa rating was measured to check removal of size. The weight loss after desizing process was 10.5 % and a Tegewa rating of 6 was observed which was due to the removal of large quantity of size present on the grey fabric.

Conventional dyeing was carried out using direct dyes on a bleached fabric, i.e. RFD (Ready for Dyeing) fabric. Also, grey, desized and desized-scoured fabrics were dyed using direct dyes by same method as of the conventional dyeing. One bath pre-treatment and dyeing was also carried out. The results in terms of dE,*K/S*, relative strength, *L*, *a*, *b*, *C*, *h* values of all the dyed fabrics are given in the Table 2. A graphical representation of relative strength (%) of all the dyed fabric samples is given in Fig. 1.

Colour strength measurement of dyed samples

Conventional dyeing is carried out on RFD fabric and thus for all comparisons the relative strength of the RFD fabric was taken as 100 %. From Table 2, it is evident that the grey fabric as well as desized fabric in all cases (dyes) showed higher relative strength and *K/S* in comparison with that of the RFD fabric. This may be attributed to the fact that starch being present on the fabric as size also has affinity for the dyes as it is chemically similar to cellulose of cotton (Figs. 2, 3). Thus, dye adsorption by the cellulose as well as starch adhered on the grey fabric gives higher adsorption of dye leading to higher *K/S* or colour strength. In case of desized fabric, although large amount of starch was removed during desizing, residual starch on desizing treatment onto the fabric which can be understood by the fact that only a weight loss of 10.5 % was observed whereas total amount of size present on the fabric was 12 %. This leads to higher absorption of dye on desized fabric compared to RFD fabric and thus higher colour strength.

Table 2. Dyeing properties of plain woven cotton fabric with direct dyes

Dyes/Samples		*dE*	*K/S*	Relative strength (%)	*L*	*a*	*b*	*C*	*h*
Red 5B	A	–	5.44	100	49.277	46.568	9.557	47.539	11.593
	B	0.70	8.57	157.46	49.686	46.473	10.128	47.564	12.289
	C	0.65	9.60	176.41	49.327	45.996	9.88	47.045	12.118
	D	0.63	7.00	128.67	49.505	47.036	9.921	48.071	11.906
	E	1.2	6.51	119.60	49.253	45.648	8.787	46.486	10.891
Blue GLL	A	–	5.35	100	41.243	−1.584	−23.681	23.734	266.139
	B	1.43	6.50	121.38	41.871	−2.859	−23.844	24.015	263.129
	C	0.78	6.38	119.10	41.606	−2.274	−23.738	23.847	264.494
	D	0.62	5.29	98.79	41.452	−2.156	−23.529	23.628	264.73
	E	1.33	6.14	114.67	41.868	−2.032	−24.772	24.855	265.276
Yellow 5GLL	A	–	7.04	100	79.484	1.974	69.327	69.355	88.333
	B	2.67	9.00	127.73	78.317	1.14	67.074	67.084	88.99
	C	3.61	8.82	125.27	77.734	1.398	66.214	66.229	88.755
	D	2.61	7.11	100.86	78.261	3.458	67.549	67.637	87.034
	E	1.93	6.87	97.56	80.533	0.657	70.272	70.275	89.428
Black FR	A	–	11.13	100	25.719	−1.222	−3.242	3.465	249.319
	B	0.32	14.08	126.44	25.584	−1.138	−2.962	3.173	248.955
	C	0.29	14.74	132.44	25.554	−1.091	−3.04	3.23	250.23
	D	0.04	11.40	102.44	25.698	−1.256	−3.256	3.49	248.878
	E	0.41	11.71	105.21	25.934	−1.554	−3.136	3.5	243.614
Olive Green BL	A	–	6.00	100	56.367	−7.205	30.301	31.146	103.406
	B	4.81	8.00	133.34	54.763	−10.358	27.032	28.949	110.993
	C	3.67	7.69	128.23	55.069	−9.492	27.739	29.318	108.919
	D	2.59	5.72	95.31	54.225	−6.397	31.53	32.172	101.5
	E	1.50	5.52	92.11	55.62	−6.141	29.537	30.169	101.776
Turquoise Blue BRLE	A	–	7.22	100	56.316	−29.568	−28.716	41.217	224.145
	B	3.75	8.04	111.35	55.412	−30.528	−25.196	39.583	219.518
	C	3.23	8.02	111.06	55.197	−30.158	−25.744	39.652	220.469
	D	1.45	6.26	86.67	55.368	−29.189	−27.676	40.224	223.458
	E	2.33	6.06	83.96	55.464	−29.778	−26.55	39.895	221.703

Where *A* control (RFD) fabric dyeing, *B* grey fabric dyeing, *C* Desized fabric dyeing, *D* Scoured fabric dyeing, *E* one bath pre-treatment and dyeing

$L = 0$ yields black and $L = 100$ indicates white, *a* negative values indicate green while positive values indicate red, *b* negative values indicate blue and positive values indicate yellow, *C* Chroma, positive values indicate More Saturated colour (i.e. brighter) and negative values indicate Less Saturated colour (i.e. duller). *h* Hue angle or purity of colour

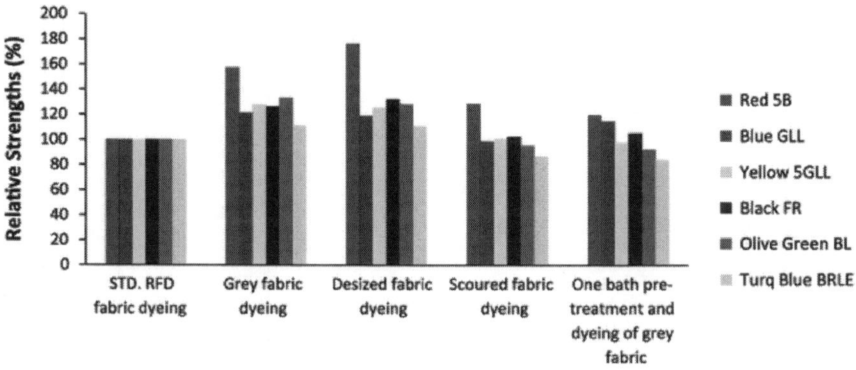

Figure 1 Relative strengths of fabrics dyed using CB Direct dyes. Relative strengths (%) of all the various dyed fabrics using direct dyes maintaining temperature of 100 °C, for 1 h using material to liquor ratio (MLR) 1:20.*Red column* indicates Red 5B dye, *dark blue column* indicates blue GLL dye, *yellow column* indicates yellow 5GLL dye, *black column* indicates black FR dye, *green column* indicates olive green BL dye and *light blue column* indicates Turquoise (Turq.) Blue BRLE dye

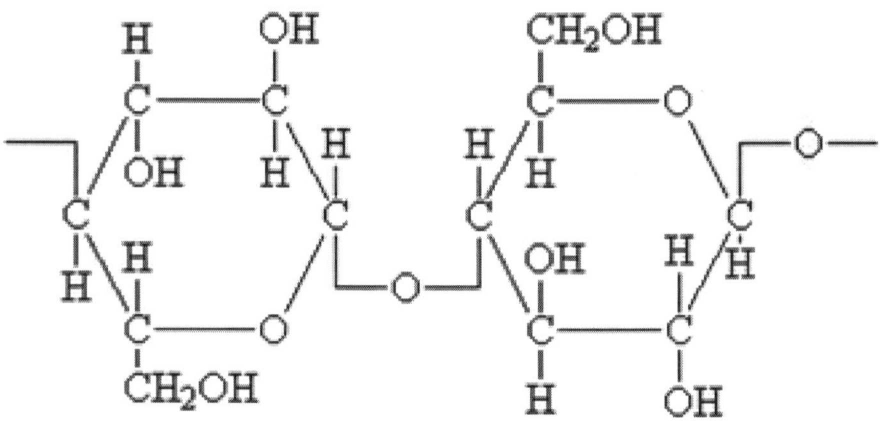

Figure 2. Structure of cellulose. Cellulose $(C_6H_{10}O_5)_n$ is a polysaccharide consisting of a linear chain of several hundred to ten thousand β-linked D-glucose units

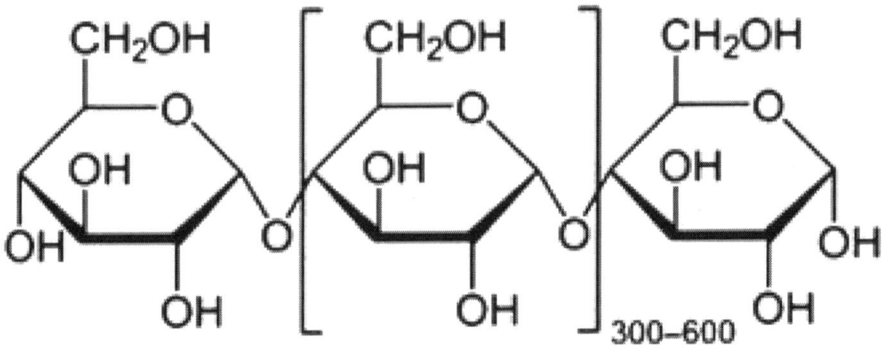

Figure 3 . Structure of starch. Starch is a carbohydrate consisting of a large number of glucose units joined by glycosidic bonds. It consists of two types of molecules: the liner (amylase) and the branched amylopectin

When scoured fabric was dyed with direct dyes, approximately same relative strength as that of conventional dyed fabric was observed. Similarly for one step pre-treatment and dyeing process, the relative strength was approximately same as the conventionally dyed fabric.

dE is a term used to denote the tonal variation between two dyed fabrics and it is observed that a $dE \leq 1$ shows a tonal variation which cannot be identified by naked eye. From Table 2, it is clear that dE for most samples has been observed to be <1 which means that there is no significant tonal variation. However, for fabrics dyed with CB Direct Olive Green BL, CB Direct Yellow 5GLL and CB Direct Turquoise Blue BRLE 150 %, a high dE has been observed which may be attributed to the fact that unbleached fabric (grey fabric) has natural coloured impurities which are leading to the tonal variation.

Table 2 also shows *L*, *a*, *b*, *C* and *h* values. Colour is judged from its lightness or darkness and its tone. The lightness or darkness is represented by the symbol L and the tone is specified by two components '*a*' and '*b*'. It can be thus interpreted from Table 2 that the *L* values of all the dyed fabrics are nearly similar to the control fabric. The only exception is with the lighter dye yellow 5GLL, where the *L* value of grey-dyed fabric is slightly less than the control. This may be due to the fact that the grey fabric is duller in appearance than the bleached fabric (control). Chroma (*C*) values of all the dyed fabrics were also approximately in line with the control fabric.

Higher relative strength of grey dyed fabric is attributed to presence of starch as size on the fabric leading to higher absorption as explained earlier. This can also hinder the penetration of the dye inside the core fibre. Therefore, it was thought appropriate to check the penetration of dyes into the interior of fibre.

For this, cross-sectional views of the dyed fabric samples were taken under optical microscope with a magnification of 50×. From Fig. 4, it can be seen that, in all the dyed samples, the dye has penetrated into the fibre similar to the control (RFD) fabric. Figure 5 shows a cross-sectional image of a denim fabric, which is ring dyed. This cross-section was taken to confirm the fact that the dyeing carried out gives well penetrated dyeing and not ring dyeing.

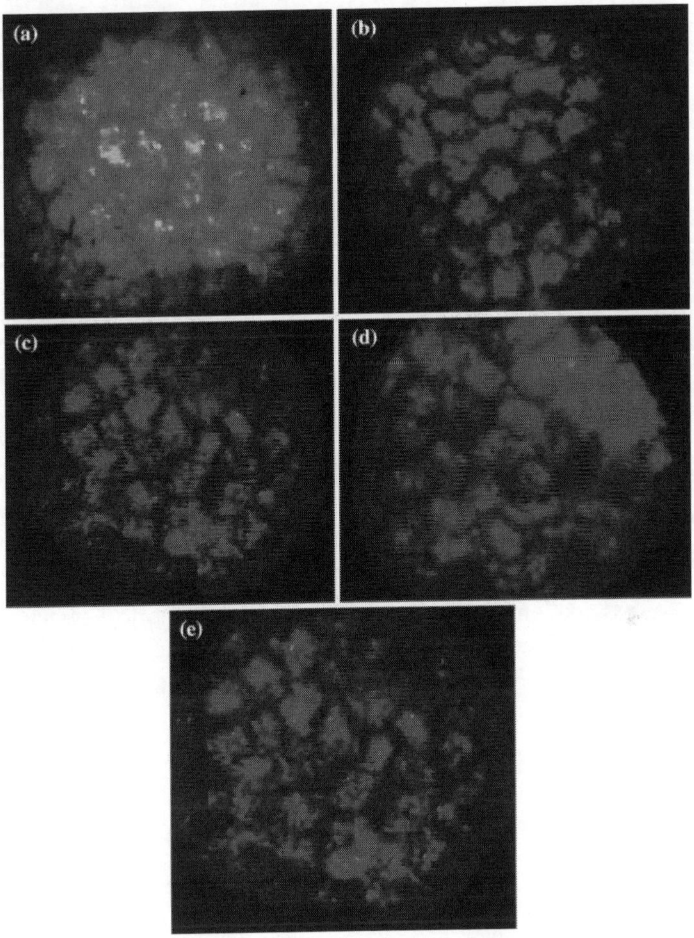

Figure 4. Cross-sectional views of yarn from direct dyed cotton fabric. Cross-sectional images of dyed yarns from control (RFD) fabric, grey dyed fabric, desized-dyed fabric, scoured-dyed fabric and one bath dyed fabric were viewed under optical microscope with a magnification of ×50. For this figure, fabrics dyed with CB Direct Blue GLL dye are chosen. Where, **a** indicates Control (RFD) fabric dyeing, **b** indicated Grey fabric dyeing, **c** indicates Desized fabric dyeing, **d** indicates Scoured fabric dyeing and **e** indicates One bath pre-treatment and dyeing of grey fabric

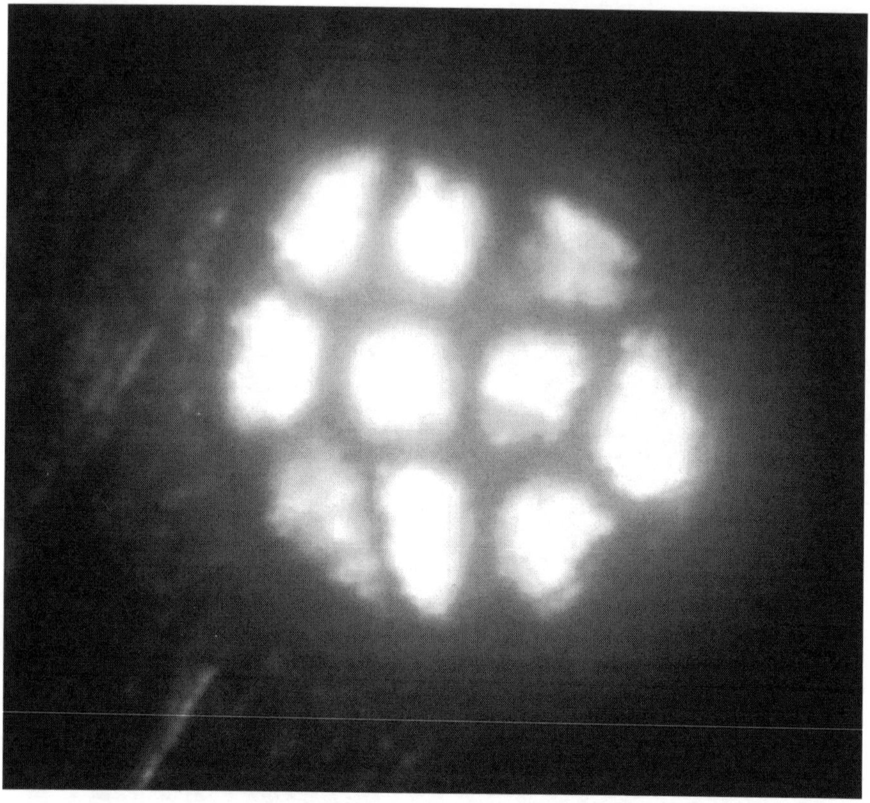

Figure 5. A cross-sectional image of a yarn from denim fabric. Few yarns form a denim fabric were removed and viewed under an optical microscope with a magnification of ×50. Ring dyeing can be seen through the cross-sectional image

Colour fastness properties of dyed samples

The fastness properties of the dyed samples were also evaluated and depicted in Table 3. All the colour fastness properties of dyed samples, i.e. washing fastness, light fastness as well as rubbing fastness were found to be at par with the control (RFD) fabric.

Table 3. Fastness properties of fabrics dyed with direct dyes

Dyes/samples		Fastness properties					
		Washing			Light	Rubbing	
		CC	SC	SW		Dry	Wet
Red 5B	A	4	1–2	3	2	4–5	4
	B	3–4	1–2	2	2	4–5	4
	C	3–4	1–2	2	2	4–5	4
	D	4	1–2	3	2	4–5	4
	E	4	1–2	2	2	4–5	4
Blue GLL	A	4–5	3–4	3	4	4–5	4
	B	4–5	3	3	3	4–5	4
	C	4–5	3	3	4	4–5	4
	D	4–5	3–4	3	4	4–5	4
	E	4–5	3–4	3–4	3	4–5	4
Yellow 5GLL	A	4–5	3–4	3	4	4–5	4–5
	B	4–5	2–3	2	4	4–5	4–5
	C	4–5	2–3	2	5	4–5	4–5
	D	3–4	2–3	3	4	4–5	4–5
	E	4	3–4	3	5	4–5	4–5
Black FR	A	4–5	4	3–4	3	4–5	4–5
	B	4–5	3–4	3	2	4–5	4–5
	C	4	3–4	3	3	4–5	4–5
	D	4	3–4	3–4	3	4–5	4–5
	E	4	3	3–4	3	4–5	4–5
Olive green BL	A	4	3	3	3	4–5	4–5
	B	4	2–3	3	2	4–5	4–5
	C	4	2–3	3	2	4–5	4–5
	D	4	2–3	3	2	4–5	4–5
	E	4	3	3	3	4–5	4–5
Turq. blue BRLE	A	4	3–4	3–4	5	4–5	4
	B	4	3	3–4	5	4–5	4
	C	4	3	3–4	5	4–5	4
	D	4	3–4	3–4	5	4–5	4
	E	4	3	4	5	4–5	4

Where *A* control (RFD) fabric dyeing, *B* grey fabric dyeing, *C* desized fabric dyeing, *D* scoured fabric dyeing, *E* one bath pre-treatment and dyeing

CC change in colour, *SC* staining on cotton, *SW* staining on wool

Absorbency of dyed fabric samples

The absorbency of the dyed fabrics was checked which can be seen in Table 4. Due to treatment at boiling temperature during one bath dyeing process, there is improvement in the absorbency as compared to the grey undyed fabric; however, it is lesser than the control fabric. Absorbency of dyed scoured fabric is similar to the control fabric as almost all the impurities are removed up to the scouring stage and only coloured components are left out to be removed further in bleaching.

Table 4. Absorbency of dyed fabrics

Samples	Absorbency (s)
Undyed RFD fabric	<3
Dyed control (RFD) sample	<3
Undyed grey fabric	>600
Dyed grey fabric	15–20
Dyed desized fabric	10–15
Dyed scoured fabric	<3
One bath pre-treatment and dyed fabric	10–12

Comparison of conventional and combined process

It is imperative to study the comparison of new process as against the well set conventional processes. Tables 5 and 6 shows the consumption of water, energy and time in conventional processes and combined process, respectively.

For calculation purpose, the fabric weight was taken as 1 kg. The total amount of water used in the conventional processes (including the washing steps) is 529 L while that required by the combined process is 89 L (as seen from Tables 5, 6). There is thus an incredible saving in the water usage which is 83.17 %.

The heat energy calculations were done using the following formula,

$$\text{Heatrequired} = m(\text{kg}) \times C_p\left(\frac{\text{J}}{\text{kg}\,^\circ\text{C}}\right) \times \Delta T(^\circ\text{C})$$

where m = mass of water (Density of water = 1,000 kg/m^3), C_p = Specific heat of water (4.186 J/kg °C)

Table 5. Water, energy and time consumption in conventional processes

Steps in the processes	Water consumption (L)	Energy consumption (kcal)	Time consumption (min)
Desizing (MLR 1:30)	30	900.4	120
Cold wash (MLR 1:50)	50	0.0	5
Hot wash (MLR 1:50)	50	3,501.7	10
Neutralisation (MLR 1:50)	50	0.0	10
Scouring (MLR 1:30)	30	2,101.0	120
Cold wash (MLR 1:50)	50	0.0	5
Hot wash (MLR 1:50)	50	3,501.7	10
Bleaching (MLR 1:30)	30	1,500.7	60
Cold wash (MLR 1:50)	50	0.0	5
Hot wash (MLR 1:50)	50	3,501.7	10
Neutralisation (MLR 1:50)	50	0.0	10
Dyeing (MLR 1:20)	19	1,330.6	60
Dye fixing treatment (MLR 1:20)	20	600.3	30
Cold wash (MLR 1:50)	50	0.0	5
Total	529	16,938	460

Table 6. Water, energy and time consumption in combined process

Steps in the processes	Water consumption (L)	Energy consumption (kcal)	Time consumption (min)
Combined pre-treatment and dyeing (MLR 1:20)	19	1,330.6	60
Dye fixing treatment (MLR 1:20)	20	600.3	5
Cold wash (MLR 1:50)	50	0.0	30
Total	89	1,930.9	95

The total heat energy required for the separate pre-treatment and dyeing processes is 16,930 kcal (70801.26 kJ); for the calculation purpose, the energy loss to atmosphere was excluded. On the other hand, the energy required for the combined pre-treatment and dyeing process is 1,930 kcal (8071.26 kJ). From this, it can be seen that there is 88.60 % energy saving in the combined process.

The total time required to process a grey fabric into a dyed one is 460 min, excluding the time required for heating and cooling of the baths. This includes all the steps of pre-treatment, i.e. desizing, scouring, bleaching and then dyeing using direct dyes. On the other hand, a combined pre-treatment and dyeing process takes only 95 min, i.e. 79.3 % reduction in time for the same work.

CONCLUSION

The single-bath pre-treatment and dyeing method saves not only the time of treatment but also water and energy. The relative strength of these dyed samples showed good results and they were very similar or in some cases slightly higher than the control sample. The nominal reduction in the relative strength observed in few cases is due to the destruction of dye molecules taking place in case of sensitive dyes due to the presence of hydrogen peroxide in the bath or the products formed during degradation of size which are of reductive nature in the bath. It can also be seen that a grey unprocessed fabric can also be dyed with acceptable properties as of the RFD fabric. The combined processes lead to substantial saving in water (83.17 %), energy (88.60 %) and time (79.3 %) along with chemicals and money.

ACKNOWLEDGMENTS

Authors wish to acknowledge the Dept. of Fibres and textile Processing Technology, ICT and Technical Education Quality Improvement Programme (TEQIP) and Fund for improvement of Science and Technology (FIST) for the Grants.

CONFLICT OF INTEREST

The authors declare that they have no competing interests.

AUTHORS' CONTRIBUTIONS

RSH has carried out all the practical work related to this research and drafted the manuscript, NRM and PBT have helped in clearing basic concepts related to the research, and helped in reviewing and drafting the manuscript. RVA has guided throughout the research work, has critically revised the manuscript and has given final approval of the version to be published. RVA has agreed to be accountable for any queries related to this manuscript. All the authors have read and approved the final manuscript.

REFERENCES

1. Zahran, M.K., Rehan, M.H., El-Rafie, M.H.: Single bath full bleaching of flax using an activated sodium chlorite/hexamethylene tetramine system. J. Nat. Fibers **2**, 49–67 (2005

2. Prabaharan, M., Rao, J.V.: Combined desizing, scouring and bleaching of cotton using ozone. Indian J. Fibre Text. Res. **28**, 444–449 (2003)

3. .Min, R.R., Huang, K.S.: Feasibility of one-step process for desizing, scouring, bleaching and mercerising of cotton fabrics: dyeing studies of direct dyes in a finite bath. J. Soc. Dyers Color **115**, 69–72 (1999)

4. Shaikh, M.A.: Water conservation in textile industry. Pakistan Text. J. **11**, 48–51 (2009)

5. Karmakar, S.R.: Chemical technology in the pre-treatment processes of textiles, 1st edn. Elsevier Sciences, Amsterdam (1999)

6. Harane, R.S., Adivarekar, R.V.: Simple approach for cost effective reuse of water in pre-treatments of cotton. Int. J. Chem. Tech. Res. **5**(2), 671–675 (2013)

7. Polonca, P., Tavcer, P.F.: Bioscouring and bleaching of cotton with pectinase enzyme and peracetic acid in one bath. Color. Technol. **124**, 36–42 (2008)

8. Saravanan, D., Sivasaravanan, S., Sudharshan Prabhu, M., Vasanthi, N.S., Senthil Raja, K., Das, A., Ramachandran, T.: One step process for desizing and bleaching of cotton fabrics using the combination of amylase and glucose oxidase enzymes. J. Appl. Polym. Sci. **123**, 2445–2450 (2012)

9. Pervin, A., Davulcu, A., Eren, H.A.: Enzymatic pre-treatment of cotton. Part 2: peroxide generation in desizing liquor and bleaching. Fibres Text. East. Eur. **73**(2), 87–90 (2009).

10. Adivarekar, R.V., Darade, N.P., Harane, R.S., Khurana, N.S.: Optimization of alkali and sodium persulphate for combined desizing-scouring-bleaching (D-S-B) exhaust Process. Colourage **59**(3), 35–41 (2012)

11. Quandt, C., Kuhl, B.: Operational possibilities and optimisation of enzymatic processes in textile finishing industry. Melliand Engl. **10**, E198–E200 (2000)

12. Hardt, P., Lund, A., May, R.: 4-in-1 process—combined bleaching and dyeing of cellulose fibres. Melliand Text. **82**, E174–E710 (2001)

13. Losonczi, A., Csiszar, E., Szakacs, G.: Bleachability and dyeing properties of biopretreated and conventionally scoured cotton fabrics. Text. Res. J. **74**, 501–508 (2004)CrossRef

14. Kokol, V., Golob, V.: The effect of single bath treatment on reactive dyeing and bio finishing of cotton fabric. In: Paper presented at the meeting of Textile Institute 83rd World Conference, Shanghai, China, 757–762 (2004)

15. Öner, E., Sahinbaskan, B.Y.: A new process of combined pre-treatment and dyeing: REST. J. Clean. Prod. **19**, 1668–1675 (2011)CrossRef

16. Chougule, M.B., Sonaje, N.P.: Novel techniques of water recycling in textile wet processing through Best Management Practices (BMP's). Int. J. Appl. Sci. Adv. Technol. **1**, 29–33 (2012)

17. AATCC: Technical Manual, vol. 75. AATCC, Research Triangle Park (1995)

18. AATCC: Test Method 110: Whiteness of Textiles. AATCC, Research Triangle Park (2005)

19. Sule, AD: Computer Colour Analysis: Textile Applications, 1st edn. New Age International Publishers, New Delhi (1997)

20. International Organization for Standardization: ISO 105C03: Textiles - Tests for Colour Fastness - Part C03: Colour Fastness to Washing. ISO, Geneva (1989)

21. International Organization for Standardization: ISO 105B02: Textiles - Tests for Colour Fastness - Part B02: Colour Fastness to Artificial Light: Xenon Arc Fading Lamp Test. ISO, Geneva (2013).

22. International Organization for Standardization: ISO 105X12: Textiles - Tests for Colour Fastness - Part X12: Colour Fastness to Rubbing. ISO, Geneva (2001)

Chapter 9

Degradation and Detoxification of Synthetic Dyes and Textile Industry Effluents By Newly Isolated *Leptosphaerulina* Sp. From Colombia

Jersson Plácido*, Xiomara Chanagá, Santiago Ortiz-Monsalve, María Yepes and Amanda Mora

Producción, Estructura y Aplicación de Biomoléculas, Universidad Nacional de Colombia Sede Medellín, Medellín, Colombia

ABSTRACT

Background

Wastewaters from the textile industry are an environmental problem for the well-known Colombian textile industry. Ligninolytic fungi and their enzymes are an option for the treatment of these wastewaters; however, the Colombian biodiversity has not been deeply evaluated for fungal strains with ligninolytic activities. In this research, 92 Colombian fungal isolates were collected from four locations around the Aburrá valley, Antioquia, Colombia. Their decolorizing activities were evaluated using Novacron Red, Remazol Black and Turquoise Blue in solid and liquid media at different culture conditions. The best fungal isolate was evaluated in the bioremediation of two real effluents and its enzymatic extracts were used in the decolorization of the three dyes.

Results

From 92 Colombian fungal isolates, *Leptosphaerulina* sp. exhibited the best decolorization percentage (>90 %) in solid and liquid cultures, and in agitated and un-agitated conditions. *Leptosphaerulina* sp. effectively decolorized the three dyes and two real effluents from textile industries. This decolorization was catalyzed by the production of significant quantities of laccase (650 U/L) and manganese peroxidase (100 U/L).*Leptosphaerulina* sp. enzymatic extracts exhibited decolorizing activity when ABTS as mediator was added. *Leptosphaerulina* sp. decolorized two real effluents from textile industries (>90 %) under conditions of low pH and glucose supplementation. Enzymatic degradation and decolorization products' innocuity was demonstrated by cytotoxic and chromatographic analyses.

Conclusion

Leptosphaerulina sp. was the best Colombian isolate. This fungal strain achieved a decolorization above 90 % for the three dyes and two real effluents from a textile industry. This decolorization was performed by producing significant amounts of laccase and manganese peroxidase.*Leptosphaerulina* sp. is an interesting prospect to treat waters polluted with dyes without the production of compounds dangerous for the environment.

Keywords: *Leptosphaerulina* sp. Ligninolytic enzymes Dyes Decolorization Textile effluents Colombia

BACKGROUND

The Colombian textile industry has national and international acknowledgements by its quality and tradition. However, this industry faces different environmental problems, one of which occurs during the fabrics dyeing. Dyeing environmental problems are related to the large demand of water and the presence of remaining dyes in their effluents. In fact, 1 kg of dyed textile requires 100 L of water generating an effluent with a dye content between 5 and 50 % (Abadulla et al. 2000). These wastewaters are disposed in rivers and creeks all around the country, generating significant pollution. Chemical dyes pollute waters by changing its color and generating harmful effects to the aquatic ecosystems. Besides the anti-aesthetic effect produced in the effluents, the coloration reduces the solar light transmittance, affecting the aquatic organisms' photosynthetic activity (Michniewicz et al. 2008). Additionally, some dyes have exhibited mutagenic and carcinogenic effects (Weisburger 2002). The traditional technologies to degrade synthetic dyes (physical, physicochemical and biological) are not efficient enough. In these technologies, the dyes are transferred to other matrices (i.e., microbial cells) or the dyes degradation products become more toxic than the original dyes (Zouari-Mechichi et al. 2006).

An environmentally friendly strategy to degrade textile dyes is the application of ligninolytic fungi and their enzymes. Ligninolytic fungi developed an enzymatic system composed by three principal enzymes: laccase (E.C 1.10.3.2), lignin peroxidase (LiP) (E.C 1.11.1.14), and manganese peroxidase (MnP) (E.C. 1.11.1.13). Laccase or benzenediol: oxygen oxidoreductase belongs to the oxidoreductases class (Dias et al. 2007). The most referenced and well-known laccases belong to the wood-decay fungi or white-rot fungi. Wood decay fungi utilize these enzymes to degrade the recalcitrant structure that constitutes lignin and facilitate the access to hemicellulose and cellulose. Laccases have the advantages of being extracellular, inducible, low substrate specificity and an inexpensive cofactor (oxygen) (Zouari-Mechichi et al. 2006; Sergio 2006; Couto and Toca-Herrera 2006). On the other hand, LiP and MnP belong to the peroxidases subclass and can catalyze the degradation of a wide number of aromatic structures, they are produced by the white rot fungi to participate in the initial states of the lignin degradation (MnP) (Angel 2002; Hammel and Cullen 2008) or the degradation of aromatic rings moderately activated (LiP) (Khindaria et al.1996). Therefore, ligninolytic fungi and their enzymes have been used in the degradation of different types of compounds as lignin, aromatic compounds, textile dyes, etc. (dos Santos et al. 2007).

Several fungi with ligninolytic activities (i.e., *Bjerkandera adusta, Trametes versicolor, Phanerochaete chrysosporium*) have proved their efficiency for treating textile dyes (dos Santos et al. 2007). However, each isolate generates a different enzymatic profile, a different response to environmental variables and different levels of dyes decolorization (Nyanhongo et al. 2002). Agitation, carbon and nitrogen source, and inducers (aromatic compounds, copper) are environmental factors that have been associated with the modification of the enzymatic production and/or the decolorization activity. However, the positive or negative effect of these factors is greatly related with the type of fungal strain evaluated. An example of this differences is the nitrogen concentration, low-nitrogen concentration has been found as favorable for*Cyathus bulleri* and *Phlebia tremellosa* (Salony et al. 2006; Robinson et al. 2001) whereas, *Pestalotiopsis* sp. and *Bjekandera adusta* showed greater enzymatic and decolorizing activity when the culture medium had nitrogen availability (Robinson et al. 2001; Hao et al. 2007).

In addition to fungal decolorization, synthetic dyes can be decolorized by using enzymatic extracts, purified or semi-purified enzymes, and commercial or native enzymes (Mattinen et al. 2011). Enzymatic decolorization have been performed by using individual or a mixture of ligninolytic enzymes, generally laccase is the main enzyme followed by MnP and LiP (Costa et al. 2005; Archibald et al. 1997). The principal advantage of enzymatic processes is the ability to obtain the same removal than microbial pretreatment in less time (between 2 and 48 h). However, some enzymes are not able to produce the same results when in vivo than in vitro. Therefore, some approaches have been utilized to increase the range of dyes degraded, the decolorization percentage, and the removal time (Zouari-Mechichi et al. 2006; Kokol et al. 2007). The principal strategy is the use of mediators. Mediators are chemical compounds, which act as electron carriers between the enzyme and the final substrate (Babot et al. 2011). The use of fungal laccases in presence of redox mediators is known as laccase-mediator system (LMS). The two most common mediators are 1-hydroxybenzotriazole (HBT) and 2,2′-azino-bis(3-ethylbenzothiazoline-6-sulphonic acid) (ABTS). A greater degree of degradation using LMS is based in a high level of interaction between the enzyme and its mediator; in fact, not all the mediators produce the same effect with enzymes from different origins (Nyanhongo et al. 2002).

Enzymatic or fungal processes using ligninolytic fungi and their enzymes are a promising technology to solve the dyes pollution generated by the Colombian textile industry. However, the fungal strains evaluated in the treatment of textiles dyes have been isolated from environmental and geographical conditions different than Colombia. Therefore, to apply this type of technology in Colombia, it is necessary to evaluate the Colombian microflora and their ligninolytic and decolorizing activities. Therefore, this research aims were the isolation of native fungi from four locations around the Aburrá valley, in Antioquia, Colombia, the evaluation of fungal strains effectiveness over synthetic dyes and actual effluents from a Colombian textile industry, and the evaluation of enzymatic extracts in the removal of synthetic dyes.

METHODS

Fungal strains

Ninety two fungi were isolated from four locations (cloud forest (CF), Montane rainforest (MRF), Medellin's Botanic Garden (BG) and an industrial and residential zone (IRZ)) around the Aburrá valley, Antioquia, Colombia. The strains were conserved in potato dextrose agar (PDA) at 4 °C until use.

Textile dyes

The dyes evaluated in this research were Novacron Red (azoic dye, λ_{max} = 514 nm), Remazol Black (azoic dye, λ_{max} = 598 nm), and Turquoise Blue (phthalocyanine dye, λ_{max} = 622 nm and 669 nm). Additionally, two types of real textile wastewaters were used. First, a dyeing bath effluent and second, the final effluent released into Medellin river. All dyes and effluents were kindly donated by Fabricato-Tejicondor S.A.

Evaluation of ligninolytic and decolorizing activity in solid media

The ligninolytic activity was measured in petri dishes with malt extract medium supplemented with 2,2′-azino-bis (3-ethylbenzothiazoline-6-sulphonic acid) also known as ABTS. The petri dish was inoculated with the isolated fungus and was incubated 2 weeks at 30 °C. The enzymatic activity was evaluated by the formation of a green or violet halo. The fungal strains that exhibited ligninolytic activity were selected for the decolorizing activity experiment. The decolorizing trials used malt extract agar supplemented with each dye. The fungi were cultivated during 20 days at 28 °C. The decolorizing activity was described as positive when the solid media exhibited partial or total loss of color. The fungal strains with decolorizing activity were selected for the liquid media evaluation.

Selection of decolorizing fungal strains

The fungal strains selection was performed in two phases. The first phase employed a completely randomized block design with three replicates. All the strains selected in the solid media experiment were evaluated in each dye. The culture was carried out in Erlenmeyer flasks with 50 mL of malt extract liquid media, 200 ppm of dye and an inoculum of four cylinders from the fungus. The fungal strains were cultured during 15 days at room temperature (20–25 °C) without agitation. At the end of the 15 days, the fungi with the largest decolorization percentage in each dye were selected for the next phase. The second phase of the selection process utilized the fungal strains selected in the unagitated culture, Kirk's media and agitation. The experiment employed a factorial design with three replicates. The experiment's factors were the selected fungi and ammonium tartrate concentration (0.22, 5, and 10 g/L). The fungal strains were cultivated during 15 days at room temperature and 150 rpm. The experiment's statistical analysis employed SAS 9.3 software. After that, a decolorization kinetic was performed for the treatments that exhibited the largest decolorization in each dye. The kinetic described the decolorization percentage, enzymatic activity, biomass, pH, protein concentration, glucose consumption, and adsorption. In order to determine the decolorization kinetic, Kirk's media was employed, with 200 ppm of dye, during 21 days at 25 °C and 150 rpm.

Selected fungus identification

The fungal isolate that presented the largest decolorization was molecular identified by analyzing the ITS and DNAr 28S sequences. The PCR used the following primers: ITS1 (5′ TCC GTA GGT GAA CCT GCG G 3′) and ITS4 (5′ TCC TCC GCT TAT TGA TAT GC 3′) and DNAr 28S LROR (5′ ACC CGC TGA ACT TAA GC 3′) and LR6 (5′ CGC CAG TTC TGC TTA CC 3′) (Chanagá Vera et al. 2012). The sequences analysis was performed in the BLAST software.

Enzymatic extracts experiment

The enzymatic extracts from the fungi with the largest enzymatic activity were evaluated in order to know their decolorizing activity. The cultures using the best fungus were stopped on the day with the largest enzymatic production. On this day, the media was centrifuged 10 min at 5000 rpm, after which, the supernatants were micro-filtrated and lyophilized. The enzymatic extracts experiment was performed for each dye, using a mixture of lyophilized extract (50 ppm), ABTS (2 mM), dye (50 ppm) and buffer pH 3. The enzymatic reaction was kept for 24 h at 25 °C and 150 rpm. The experiment had three replicates and the decolorization percentage as response variable. The statistical analysis was performed using the SAS 9.3 software.

Fungal decolorization of real textile wastewaters

The fungus with the greatest decolorization ability was evaluated with two types of textile wastewaters (the final effluent disposed into Medellin river and a fabrics dye bath). The final effluent was evaluated under a completely randomized factorial design 2 × 2 × 2 with three replicates and using the decolorization percentage as response variable. The experiment's factors were medium supplementation (10 g/L of glucose and 5 g/L of ammonium tartrate), pH (as received and pH 5) and inoculum (4 and 8 cylinders). The fabrics dyeing bath was evaluated under a completely randomized factorial design 2 × 2 × 2, using as factors medium supplementation (10 g/L of glucose and 5 g/L of ammonium tartrate), inoculum (4 and 8 cylinders) and effluent dilution [as received (6.8 g/L) and 200 ppm]. The decolorization was carried out in 250 mL Erlenmeyer flasks with 50 mL of medium, during 21 days at 25 °C and 150 rpm.

Enzymatic activities

The laccase activity was calculated using the enzymatic oxidation of ABTS. The reaction was followed spectrophotometrically by employing the change in the absorbance at 420 nm and a molar extinction coefficient of ε_{420}, 36,000/M/cm. The LiP activity was obtained by following the formation of veratryl aldehyde from veratryl alcohol at 310 nm (ε_{310}, 9300/M/cm). The MnP activity was calculated by measuring the absorbance change at 469 nm produced by the oxidation of 2,6-dimethoxypenol (DMF) (ε_{469}, 27,500/M/cm). All the enzymatic activities were reported in units (U), which were defined as the quantity of enzyme that catalyze 1 µmol of substrate per minute.

Analytical methods

The decolorizing activity was calculated using the decolorization percentage decolorization%=A0−AfA0×100decolorization%=A0−AfA0×100, where A_0 is the initial dye absorbance at the dye's λ_{max} and A_f is the final absorbance at the dye's λ_{max} (Mohorčič et al. 2006). The decolorizing activity was considered high when the decolorization percentage was above 80 %, medium when it was between 50 and 80 % and low when it was below 50 %. The error bars utilized in the decolorization percentage measurements corresponds to the standard error. To evaluate biomass adsorption, the biomass was re-suspended in methanol for 30 min, and the methanol solution was used to calculate the absorbance at the dyeλ_{max}. In order to obtain the biomass production, after the methanol extraction, the biomass was dried at 105 °C for 24 h until obtain constant weight. The glucose determination was performed using the 3,5-dinitrosalicylic acid (DNS) methodology. The protein concentration was obtained by the Bradford's method. Dyes' degradation products were analyzed by thin layer chromatography (TLC) and 3-(4,5-Dimethylthiazol-2-yl)-2,5-diphenyltetrazolium bromide (MTT) cytotoxicity test. Lyophilized extracts from fungal cultures were employed for both analyses. In the TLC, the lyophilized extracts were solubilized in different organic solvents (methanol, acetonitrile, ethyl acetate, and hexane). Each organic extract was spotted in the solid phase (silica gel U60 plates) and its respective extraction solvent was used as mobile phase. The degradation-products toxicity was evaluated using the MTT test. This test evaluated the effect of the lyophilized extracts in the cell-viability percentage of human cell line U937 (Gallego et al. 2006). The cytotoxicity was measured by comparing the absorbance (595 nm) produced by the control cells and the absorbance produced by the cells exposed to different concentration of the lyophilized extracts (500, 250, 125, 62.5, 31.2, 15.6, and 7.81 ppm). A compound was considered toxic, when the absorbance of the cells exposed to the compound exhibited an absorbance lower than the control's absorbance (Gallego et al. 2006).

RESULTS

Evaluation of ligninolytic and decolorizing activity in solid media

In total, 92 fungi were collected from the four isolation locations. The strains were obtained from different types of lignocellulosic material (composted biomass, trees, discomposing trees etc.). The isolates distribution by site was IRZ (15), BG (32), CF (25), and MRF (20). After isolation and strains purification, the fungi were inoculated in solid malt extract media supplemented with ABTS. In solid media was possible to observe green and purple halos and a combination of both halos. The purple halos were observed after the production of small green halos. The halo formation fluctuated from hours (12–24 h) to days (2–12 days). In total, 30 of 92 isolates exhibited ligninolytic activity. The isolates ID and isolation location are described in Table 1. The total percentage of fungi with enzymatic activity was 32 % and the percentages of positive fungi by sites were IRZ (60 %), BG (15.6 %), CF (40 %), and MRF (32 %).

Table 1 .Fungal isolate and isolation location of the fungal strains with ligninolytic activity

Isolate	Isolation location	Isolate ID	Isolation location
LVN 4	Botanic garden	LVN 63	Cloud forest
LVN 5	Botanic garden	LVN 65	Cloud forest
LVN 6	Botanic garden	LVN 71	Cloud forest
LVN 31	Botanic garden	LVN 72	Cloud forest
LVN 32	Botanic garden	LVN 73	Cloud forest
LVN 35	Montane rainforest	LVN 74	Cloud forest
LVN 42	Montane rainforest	LVN 79	Industrial and residential zone
LVN 44	Montane rainforest	LVN 80	Industrial and residential zone
LVN 45	Montane rainforest	LVN 82	Industrial and residential zone
LVN 48	Montane rainforest	LVN 83	Industrial and residential zone
LVN 49	Montane rainforest	LVN 84	Industrial and residential zone
LVN 51	Montane rainforest	LVN 85	Industrial and residential zone
LVN 52	Montane rainforest	LVN 88	Industrial and residential zone
LVN 54	Cloud forest	LVN 91	Industrial and residential zone
LVN 61	Cloud forest	LVN 92	Industrial and residential zone

The 30 selected isolates were evaluated in malt extract agar with each of the three dyes (Novacron red, Remazol black, and Turquoise blue). In this case, an isolate was classified as positive for decolorizing activity when the media exhibited partial or total decolorization. In total, 15 fungi exhibited decolorizing activity which corresponds to 50 % of the preselected fungi and 16.3 % of the original isolated fungi. The percentage of positive isolates was relatively high. In fact, other reports with more than 50 isolates obtained 5.7 % (Hernández-Luna et al.2008) and 10 % (Risna and Suhirman 2002) of fungi with decolorizing activity. The fungal isolates with decolorizing activity and their level of decolorization are reported in Table 2. Remazol black was degraded by 12 strains, two of which (LVN6 and LVN63) showed the greatest decolorization. Additionally, LVN6 and LVN63 obtained the fastest decolorization with 5 and 3 days, respectively. Similar decolorization velocity have been reported in *P. chrysosporium* (Máximo et al. 2003), *T. versicolor* (Mohorčič et al. 2006), *Bjerkandera adusta* and *T. versicolor* (Novotný et al. 2004). In contrast, the other fungal strains took from 8 to 15 days in decolorize Remazol Black. On the other hand, Turquoise blue obtained the largest number of fungal strains with decolorization (13 strains); however, the decolorization realized by the fungal strains had medium or low level (Table 2). In Turquoise Blue, the greatest decolorization was obtained after day 15. Opposite to Turquoise Blue, Novacron Red obtained the lowest number of fungal isolates (4) with decolorizing activity. This low quantity of positive strains was related to the dye's monoazoic structure. Monoazoic dyes have been reported in different studies to exhibit a high difficulty to be degraded by ligninolytic fungi. This difficulty is associated with the enzyme access to the dye's azo group (–N = N–) (Máximo et al. 2003; Park et al. 2007).

Table 2. Qualitative evaluation of the decolorizing ability in the selected fungal strains

Isolate	Dye		
	Turquoise Blue	Remazol Black	Novacron Red
LVN 6	Medium	High	Medium
LVN 35	Low	Low	Absence
LVN 44	Low	Low	Absence
LVN 45	Low	Low	Absence
LVN 48	Low	Absence	Absence
LVN 49	Medium	Absence	Absence
LVN 52	Absence	Medium	Absence
LVN 54	Medium	Low	Absence
LVN 61	Medium	Absence	Absence
LVN 63	Low	High	Medium
LVN 71	Low	Low	Absence
LVN 72	Low	Absence	Absence
LVN 74	Low	Medium	Absence
LVN 80	Low	Medium	Medium
LVN 88	Absence	Medium	Low

Selection of decolorizing fungal strains

The fungal strains selection was performed in two phases. The first phase was a preselection study employing malt extract medium without agitation. Whereas, the second phase was an experiment using agitation, Kirk's medium and variations in the nitrogen concentration.

Fungal strains preselection

The preselection experiment was carried out with the 15 isolates that exhibited decolorization activity in solid media (Fig. 1). This experiment evidenced that all the strains generated decolorization; however, the removal level varied among strains and dyes. The variation between solid and liquid media is explained by the changes in environmental conditions, it has been found that fungi behaves different between solid and submerged cultures, in fact, enzymes production is favored by growing in solid surfaces improving the enzymatic activity and the enzymatic profile (Novotný et al. 2004). In Remazol Black, the isolates LVN6, LVN88, and LVN63 exhibited the highest removal (>80 %), whereas the other fungi removal percentages varied between 20 and 80 %. The results between solid and liquid experiments were similar in LVN6 and LVN63; in both, they achieved the largest decolorization. In Novacron Red, the fungal isolates LVN80, LVN61, LVN6, LVN63 and LVN88 exhibited the most significant decolorization. From these fungi, LVN61 was the only strain with high decolorization in the liquid experiment and low decolorization in the solid experiment. The high removal observed in Novacron Red, under non-agitated conditions, is not common. In fact, monoazoic dyes had low or none decolorization under non-agitated conditions (Jarosz-Wilkołazka et al. 2002; Swamy and Ramsay 1999). Turning now to turquoise blue, five strains (LVN6, LVN49, LVN63, LVN80, and LVN72) obtained decolorization around 70 % or more. This result is promissory because under non-agitation conditions other phthalocyanines dyes have not been decolorized by other fungal strains (Swamy and Ramsay 1999).

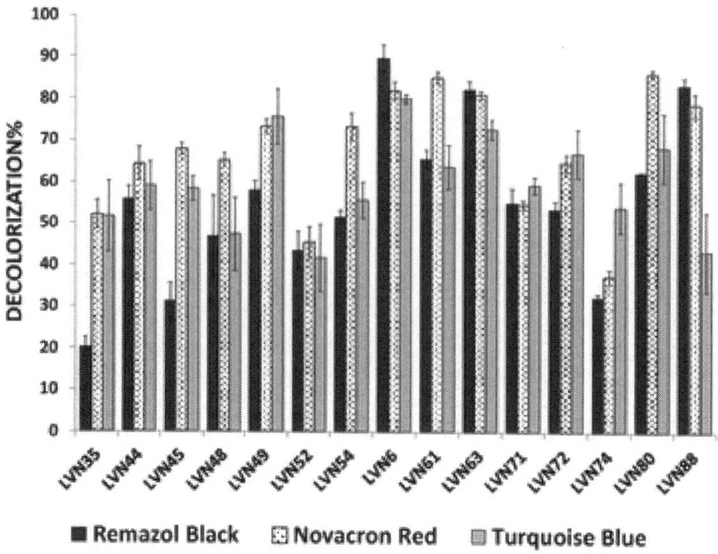

Figure 1. *Barplot* of the decolorization percentage obtained by the fungal strains in the preselection experiment without agitation and malt extract media

The experiment's statistical analysis (see Additional file 1: Annex S1) was based in the decolorization percentage, using the Duncan's test to select the best fungal isolates for each dye (Table 3). The isolates selected were LVN6, LVN63, LVN80, LVN88, LVN61, LVN49, and LVN72. The decolorization percentage of the selected fungal strains were around 85 % (see Fig. 1). These values were similar to the values achieved by cultures without agitation of *Phlebia tremellosa, Irpex lacteus,* and basidiomycete strain PV 002 (Kirby et al. 2000; Svobodová et al. 2008). The isolates LVN6 and LVN63 were the only strains with high removal in all the dyes evaluated. The ability of decolorize different types of dyes, under non-agitation conditions, was reported in a small number of fungal strains (Kirby et al. 2000; Erkurt et al. 2007).

Fungal strains selection

The fungal strains selection was performed in Kirk's medium with agitation and three different concentrations of ammonium tartrate. The fungal strains used in each dye are shown in Table 3. The experiment employed a factorial design, using as factors the selected microorganism for each dye and the three concentrations of ammonium tartrate. The factorial design produced nine treatments for Remazol Black and 15 treatments for Novacron Red and Turquoise Blue. The decolorization percentages obtained by each treatment are shown in Fig. 2.

Table 3. Preselected fungal isolates

Turquoise Blue	Remazol Black	Novacron Red
LVN 6	LVN 6	LVN 6
LVN 63	LVN 63	LVN 63
LVN 80	LVN 88	LVN 80
LVN 72		LVN 88
LVN 49		LVN 61

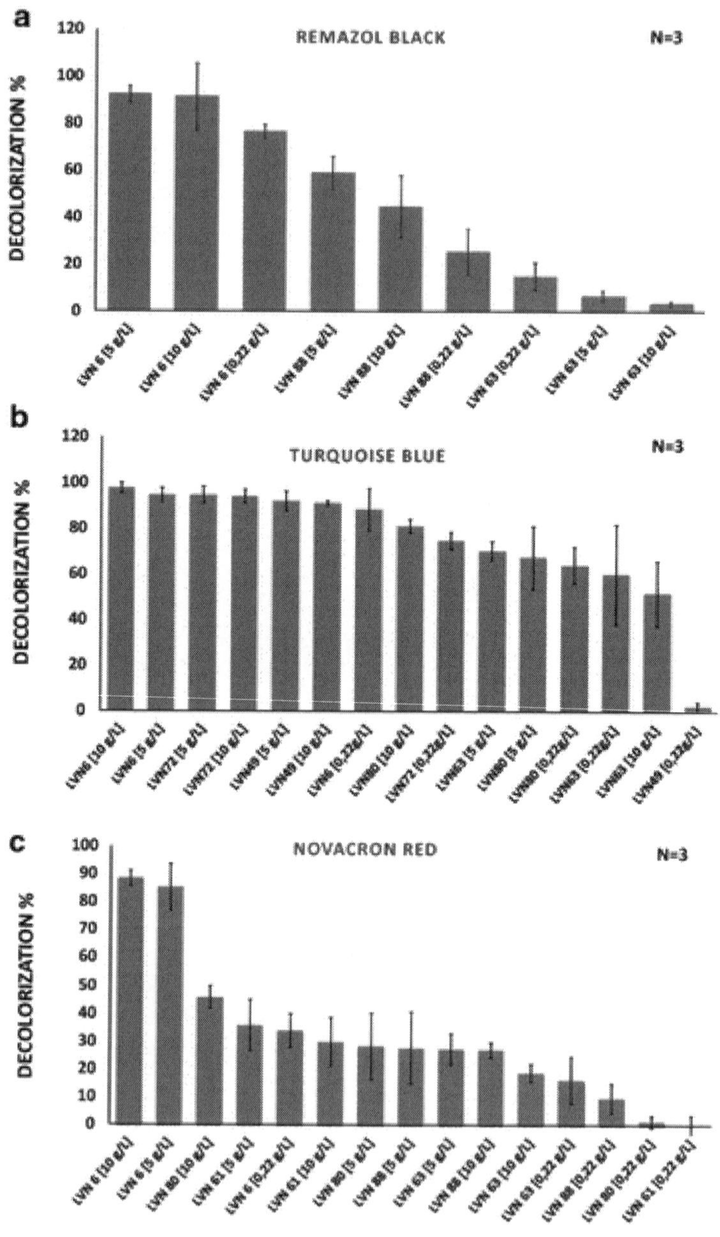

Figure 2. *Barplot* of the decolorization percentage obtained in the selection experiment with Kirk's media and agitation; **a** Remazol Black, **b**Turquoise Blue, **c** Novacron Red

Remazol Black experiments showed LVN6 as the best fungal strain (>90 %). In this dye, LVN6 at the three concentrations of ammonium tartrate realized decolorization greater than the other fungal strains (Fig. 2a). The statistical analysis showed interaction between the factors; therefore, the treatments selection was performed using the interaction plots (see Additional file 1: Annex S1). The interaction plots evidenced the greatest decolorization in the treatments with LVN6 and 5 and 10 g/L of ammonium tartrate. In contrast, LVN63 obtained the highest removal at 0.22 g/L of ammonium tartrate; this behavior was also detected in Novacron Red. The decolorizing activity observed in LVN6 [5 g/L] and LVN6 [10 g/L] was similar to *Bjerkandera adusta, Trametes versicolor, Funalia trogii* and *Irpex Lacteus,* strains recognized by their high decolorizing activity over Remazol Black (Mohorčič et al. 2006; Park et al. 2007; Máximo and Costa-Ferreira 2004). In Remazol Black, the selected treatment was LVN6 [5 g/L] because it produced the highest decolorization with the lowest consumption of ammonium tartrate.

The highest removals were achieved in Turquoise Blue. In this dye, almost all the treatments presented removals above 50 %. In this dye, three strains (LVN6, LVN72 and LVN49) obtained decolorizations above 90 % (Fig. 2b). The decolorization obtained by LVN6, LVN72 and LVN49 is comparable with the removal obtained by *P. simplicissimum* INCQS40211, *P. chrysosporium,* and *Ph. Tremellosa* over Turquoise Blue. In these strains, the removal was above 94 % (Kirby et al. 2000; Bergsten-Torralba et al. 2009). Similar to Remazol black, the statistical analysis evidenced the interaction between isolates and ammonium tartrate concentration. The interaction plots indicated LVN49, LVN72 and LVN6 as the best fungal strains to decolorize Turquoise Blue and the concentrations of 5 and 10 g/L of ammonium tartrate as the best concentrations to obtain high decolorization. In this dye, the best treatments were LVN49 [5 g/L], LVN72 [5 g/L], and LVN6 [5 g/L]. However, the treatments LVN49 [5 g/L] and LVN72 [5 g/L] were not employed in further experiments because its lack of efficiency over the other two dyes.

Novacron Red experiment exhibited LVN6 [5 g/L] and LVN6 [10 g/L] as the treatments with the greatest decolorization percentages (>85 %) (Fig. 2c). On contrary, the other fungal strains had a reduction in the decolorizing ability, in comparison with the unagitated experiment. This reduction can be explained by the changes in the growth media and the inclusion of agitation. The statistical analysis (see Additional file 1: Annex S1) exhibited significant differences among the treatments and an interaction between the factors. The interaction plots revealed 10 and 5 g/L of ammonium tartrate and LVN6 as the best treatments for Novacron Red decolorization. The removal obtained by LVN6 [5 g/L] (85.3 %) and LVN6 [10 g/L] (88.5 %) was larger than the reported by other fungal strains at similar or greater concentrations (≥200 ppm) of monoazoic dyes (Jarosz-Wilkołazka et al. 2002). The treatment LVN6 [5 g/L] was selected as the best option for the decolorization of Novacron Red, because it reached a similar decolorization than LVN6 [10 g/L], using half of the concentration of ammonium tartrate.

Molecular identification of the isolate LVN6

The ITS region (DNAr) and DNAr 28S from LVN6 were amplified to identify the molecular identity of this strain. The amplification of these DNA sequences generated bands of 500–600 pair of bases (pb) for ITS, and 1100 pb for DNAr 28S. The BLAST analysis and sequence identity of ITS and DNAr 28S identified the fungi LVN6 as a member of the genre *Leptosphaerulina*. A specie determination using the blast analysis was not possible. In this strain, the sequences of the ITS region and DNAr 28S were homologous to two different species of *Leptosphaerulina*. The ITS region's sequence was homologue to *Leptosphaerulina Americana*, whereas DNAr 28S' sequence was homologous to *L. australis.* The phylogenetic and morphological analysis of these sequences did not allow the complete identification of this strain (Chanagá Vera et al. 2012). At this moment, this is the first report demonstrating the decolorizing activity in the genre *Leptosphaerulina.* The ability of *Leptosphaerulina* sp. to decolorize solutions with dyes can be related to the presence of ligninolytic enzymes. Likewise, *L. chartarum* and *L. briosiana* have been described as ligninolytic enzymes producers (Simon et al. 1979; Sajben-Nagy et al. 2014).

Decolorization kinetics

Figure 3a, b shown the Remazol Black decolorization kinetic. In the first 5 days, Remazol Black decolorization was approximately 30 %; however, after seven days, it increased above 90 %. The decolorization was correlated with the increment in the enzymatic production (Fig. 3a). *Leptosphaerulina* sp. produced laccase and MnP during all the kinetic. In Remazol Black, the greatest production of enzymes started at day 7 and had its maximum production around day 15. The highest production of laccase and MnP was 650 U/L and 100 U/L, respectively. Turquoise Blue decolorization kinetic exhibited a removal above 90 % in 3 days (Fig. 3c, d). The initial removal was performed by an adsorption process. The Turquoise Blue high adsorption can be related to the copper molecule in the core of the dye structure. After the day 5, the adsorption process was replaced by the enzymatic process. The enzymatic catalysis removed the dye particles attached to the fungal biomass. This removal was catalyzed by laccase and MnP production. The largest enzymatic production was observed at day 9 with a laccase activity of 512 U/L and a MnP activity of 100 U/L. The Novacron Red decolorization kinetic by *Leptosphaerulina* sp. [5 g/L] (Fig. 3e, f) showed a removal greater than 90 % and the production of ligninolytic enzymes. The highest decolorization achieved by *Leptosphaerulina* sp. was obtained in 9 days, the enzymatic production was observed during all the kinetic; nevertheless, the highest enzymatic activities of laccase and MnP were observed on the 30th day with 647 U/L and 90 U/L, respectively.

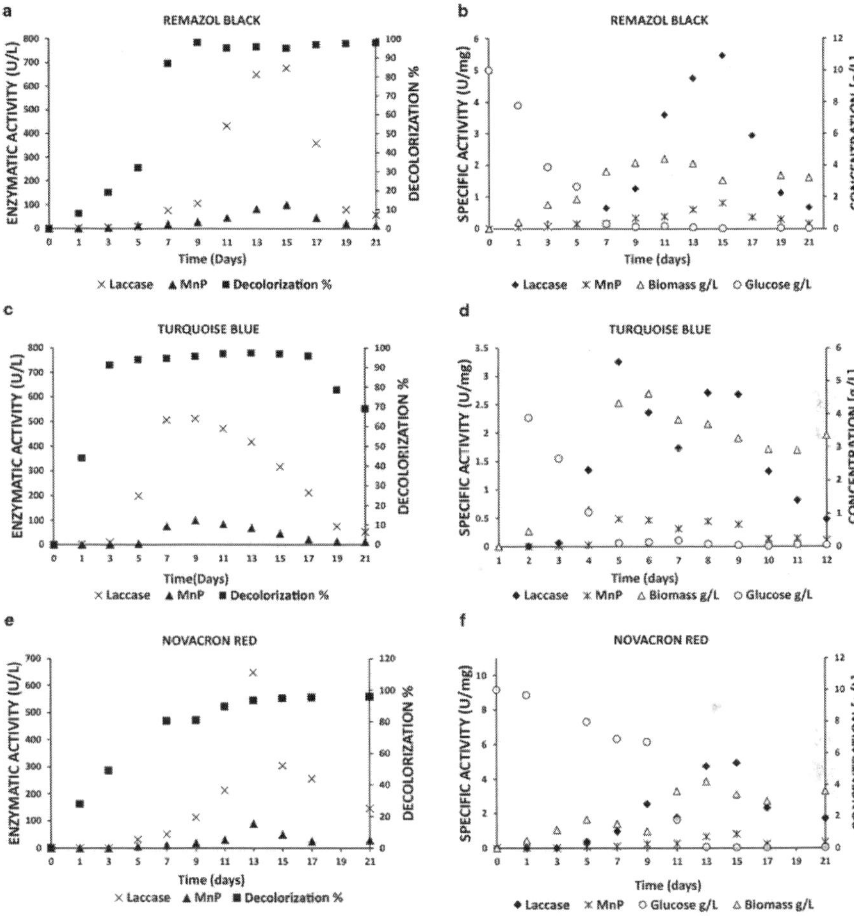

Figure 3. *Leptosphaerulina* sp. decolorization kinetic. **a** Remazol Black kinetic showing decolorization % and enzymatic activities, **b** Remazol Black kinetic showing glucose concentration, biomass concentration, and specific enzymatic activities **c**Turquoise Blue kinetic showing decolorization percentage and enzymatic activities, **d** Turquoise Blue kinetic showing kinetic showing glucose concentration, biomass concentration, and specific enzymatic activities **e** Novacron Red showing decolorization percentage and enzymatic activities, **f**Novacron Red showing showing kinetic showing glucose concentration, biomass concentration, and specific enzymatic activities

Enzymatic decolorization using *Leptosphaerulina* sp. extracts

The enzymatic degradation of the three dyes was evaluated by employing a commercial laccase and ABTS as mediator (see Additional file 1: Annex S1). The in vitro analysis showed that all the dyes were decolorized by the commercial laccase; however, the level of decolorization varied depending on the dye and ABTS concentrations. The highest decolorization was obtained in Turquoise Blue (83 %), followed by Remazol Black (80 %), and Novacron Red (56 %). The greatest decolorizations were achieved by the treatments with ABTS as mediator (2 mM) and the highest commercial enzyme concentration (2 mg/ml). Turquoise Blue was the only dye that exhibited any level of decolorization without using ABTS. The *Leptosphaerulina* sp. lyophilized-extracts (Fig. 4) decolorized enzymatically all the dyes. In Remazol Black, the enzymatic decolorization was approximately 70 %. *Leptosphaerulina* sp. enzymatic extract transformed the reaction media color, from a deep black into a reddish color. *Leptosphaerulina* sp. enzymatic extracts achieved similar levels of decolorization than the commercial enzyme (80 %). In Turquoise Blue, the *Leptosphaerulina* sp. enzymatic extracts produced a decolorization of 56 % (Fig. 4). In Turquoise Blue, the decolorization difference between the commercial laccase (83 %) and *Leptosphaerulina* sp. enzymatic extracts (56 %) evidenced a necessity to improve the enzyme purification process. In Novacron Red, the enzymatic extracts achieved low removal (20 %); nevertheless, the commercial enzyme also got low removal (56 %).

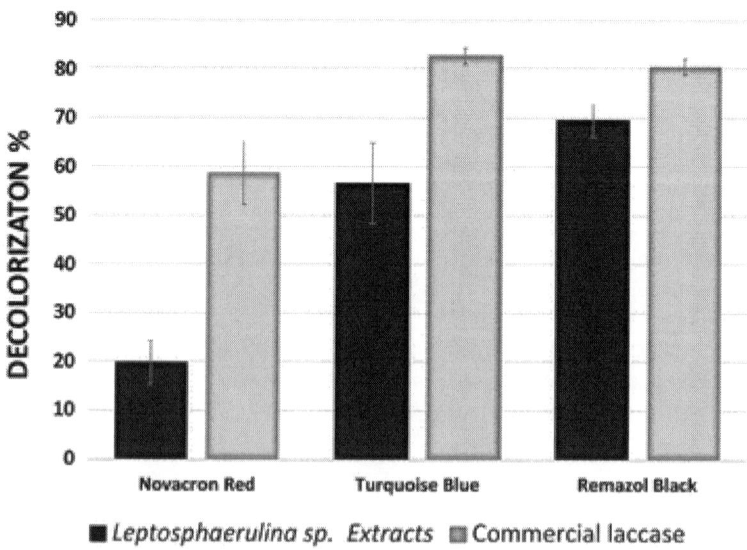

Figure 4.*Barplot* of the decolorization percentage obtained by the in vitro study of *Leptosphaerulina* sp. enzymatic extracts

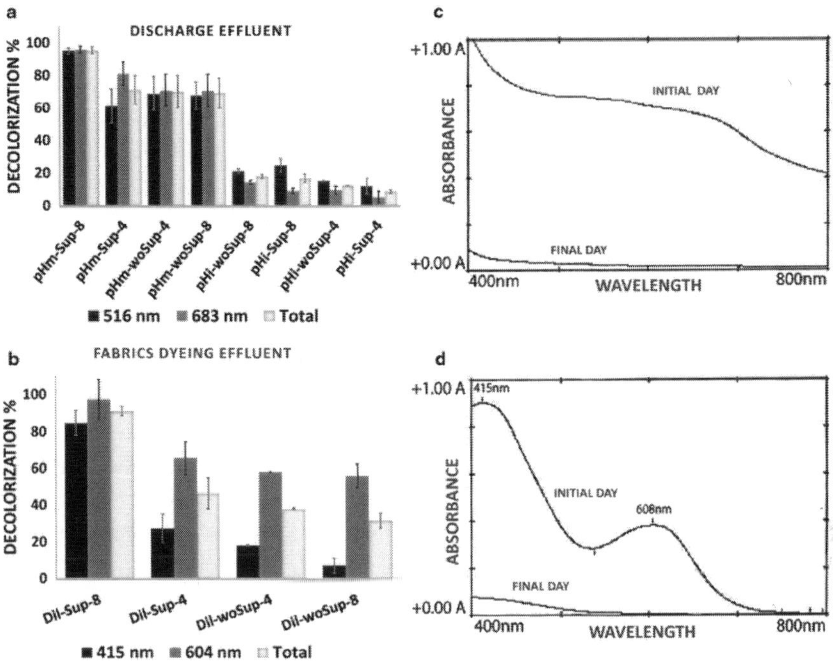

Figure 5. Real textile industry effluents decolorization by *Leptosphaerulina* sp. **a** *Barplot* of the discharge effluent experiment. **b** *Barplot* of the dyeing effluent experiment. **c** Discharge effluent spectra from pHm-Sup-8 treatment at the initial and final day. **d** Dyeing effluent spectra Dil-Sup-8 treatment at the initial and final day

Real textile effluents treatment using *Leptosphaerulina* sp.

The final discharge effluent had the following characteristics: alkaline pH (12), greyish color, low dyes concentration, and the presence of suspended particles. The effluent spectrophotometric analysis showed two strong signals at 520 and 680 nm. These signals were employed to calculate the effluent decolorization percentage. The decolorizing effect of *Leptosphaerulina* sp. was evaluated using a factorial design 2 × 2 × 2. The experiment's factors were: medium supplementation (10 g/L of glucose and 5 g/L of ammonium tartrate), pH (as received and pH 5) and inoculum (4 and 8 cylinders). After 21 days, the treatments with low pH achieved a decolorization above 70 %, whereas, the treatments with the original pH produced removals below 30 % (Fig. 5a). The experiment's statistical analysis evidenced a lack of interaction between the three factors (*p* value >0.05). Thus, the best treatment selection was carried out by doing the Duncan's test (see Additional file 1: Annex S1). The treatment pHm-Sup-8 achieved the best decolorization (95.5 %), followed by pHm-Sup-4

(71 %), pHm-woSup-4 (70 %), and pHm-woSup-8 (69 %). The treatments that produced the largest decolorization also showed laccase production. pHm-Sup-8 treatment achieved the greatest laccase production (24.86 U/L), followed by pHm-Sup-4 (5.27 U/L). The first and final days effluent's spectrum evidenced the complete reduction of the signals observed at the beginning of the experiment (Fig. 5c).

The dyeing effluent was evaluated using a factorial design 2 × 2 × 2. In this effluent, the experiment's factors were: medium supplementation (10 g/L of glucose and 5 g/L of ammonium tartrate), inoculum (4 or 8 cylinders), and dilution (as received and 200 ppm of dyes, the dye concentration was calculated based in Novacron Red's original concentration). The effluent had three dyes: Novacron Red (6.8 g/L), Novacron Blue (6.75 g/L), and Novacron Yellow (5.94 g/L). The effluent had a basic pH (12), suspended particles and deep military green coloration. The spectrophotometric analysis revealed two strong signals at 415 and 608 nm (Fig. 5d); these signals were used to calculate the decolorization percentage. *Leptosphaerulina* sp. only decolorized the effluent when it was diluted to 200 ppm (Fig. 5b), the undiluted samples did not show decolorization. The largest removal was obtained by Dil-Sup-8 (90 %), the other diluted treatments decolorized the effluent below 50 %. Dil-Sup-8 exhibited a considerable reduction in the signal at 415 nm (Fig. 5d). The absorbance reduction in this peak can be related with the production of laccases. In fact, Dil-Sup-8 presented the most significant laccase production (40 U/L). The absorbance reduction at 415 nm was observed in the treatment Dil-Sup-4, which also produced laccase activity (10 U/L). The statistical analysis evidenced an interaction between the inoculum quantity and the medium supplementation (see Additional file 1: Annex S1). The interactions plot described an inverse interaction between the factors. Additionally, the interaction plot confirmed the treatment Dil-Sup-8 as the best conditions to decolorize the dyeing effluent by *Leptosphaerulina* sp.

Decolorization products evaluation

The decolorized media (dyes and effluents) were lyophilized; then, the compounds in the lyophilized were extracted using four solvents (methanol, acetonitrile, ethyl acetate, and hexane). After that, the thin layer chromatography (TLC) was performed. The extracts were spotted and run it using the same solvent as mobile phase. The hexane extracts did not produce any type of signals under UV light (near and far) and visible light. By contrast, the methanol, acetonitrile and ethyl acetate extracts revealed signals under UV and visible lights (Table 4). The chromatograms of the three dyes and the effluents exhibited the disappearance of the signals corresponding to the dyes were present in dissolution and effluents (Rf 1). Additionally, the methanol, acetonitrile and ethyl acetate extracts showed a large number of signals. These signals were not observed in the controls (Kirk's media with and without dye). Turquoise Blue extracts generated a number of signals greater than Remazol Black and Novacron Red extracts. The use of TLC to evaluate the fungal or enzymatic degradation of dye have been reported in the enzymatic decolorization of Reactive Orange 16, Reactive Red 2 and Reactive Blue 4 (Axelsson et al. 2006; Telke et al. 2009). In these publications, the disappearance of the original dye's signal and the appearance of one or more signals with different Rf were correlated with fungal or enzymatic dye degradation. Using the same proposition, *Leptosphaerulina* sp. degraded the dyes in Kirk's media and the real effluents. However, it was not possible to identify how and which of the signals were related to the dye degradation.

Table 4. Thin layer chromatography results of the lyophilized extracts obtained after the decolorization process

Dye/effluent	Solvent	Number of new bands			
		Visible	Near UV	Distant UV	Total
Remazol Black	Methanol	3 (Rf: 0.72, 0.78, 0.86)	3 (Rf: 0.72, 0.78, 0.86)	3 (Rf: 0.52, 0.72, 0.8)	9
	Acetonitrile	0	2 (Rf: band and 0.84)	2 (Rf: 1, 0.72)	4
	Ethyl acetate	0	1 (Rf: band between 0.70 and 0.79)	4 (Rf: 0.6,0.71, 0.79, 0.83)	5
Novacron red	Methanol	1 (Rf: 0.69)	3 (Rf: 0.67, band between 0.7 and 0.74, 0.79)	4 (Rf: 0.55, 0.75, 0.78, 0.87)	8
	Acetonitrile	0	3 (Rf: band, 0.52, 0.86)	3 (Rf: 0.68, 0.91, 1)	6
	Ethyl acetate	0	3 (Rf: 0.25, 0.31, 0.51, 0.75)	4 (Rf: 0.62, 0.79, 0.81, 0.85)	7
Turquoise blue	Methanol	1 (Rf: 0.69)	3 (Rf: 0.69, band between 0.71 and 0.77, 0.82)	4 (Rf: 0.55, 0.67, 0.7, 0.8)	8
	Acetonitrile	0	2 (Rf: barrido y 0.88)	5 (Rf: 0.76, 0.79, 0.83, 0.93, 1)	7
	Ethyl acetate	0	5 (Rf: 0.13, 0.19, 0.25, and 0.33 and a band between 0.66 and 0.8)	5 (Rf: 0.74, 0.78 0.81, 0.86, 0.9)	10
Discharge effluent	Methanol	0	2 (Rf: 0.65, 0.76)	1 (Rf: 0.65)	2
	Acetonitrile	0	1 (Rf: 1)	0	1
	Ethyl acetate	0	2 (band and Rf: 0.7)	6 (Rfs: 0.09, 0.16, 0.22, 0.5, 0.72, 0.84)	8
Dyeing effluent	Methanol	1 (Rf: 0.65)	1 (Rf: 0.65)	1 (Rf: 0.65)	1
	Acetonitrile	0	1 (Rf: 0.78)	0	1
	Ethyl acetate	0	3 (Rf: 0.65, 0.1, 0.2)	2 (Rf: 0.71, 0.76)	5

The MTT test used lyophilized extracts from decolorized cultures of *Leptosphaerulina* sp. The MTT analyzed the cytotoxic effect of different concentrations of the lyophilized extracts over the human cell line U937. At the end of the test, the concentrations evaluated did not evidence cytotoxicity to the human cell line U937, because none of the concentrations generated a significant reduction in the absorbance at 595 nm (Table 5). These results evidenced that *Leptosphaerulina* sp. decolorized Novacron Red, Remazol Black and Turquoise Blue, without the production of toxic compounds.

Table 5. Cytotoxicity analysis of the lyophilized extracts obtained after the decolorization process using the MTT test

Lyophilized extracts concentration (ppm)	Remazol Black Absorbance	Novacron Red Absorbance	Turquoise Blue Absorbance
500	1.09 ± 0.07	1.03 ± 0.04	1.02 ± 0.09
250	1.11 ± 0.10	0.68 ± 0.06	0.68 ± 0.10
125	0.99 ± 0.25	0.79 ± 0.05	0.68 ± 0.05
62.5	1.01 ± 0.05	0.81 ± 0.01	0.73 ± 0.07
31.2	1.04 ± 0.10	0.77 ± 0.02	0.68 ± 0.04
15.6	1.03 ± 0.05	0.64 ± 0.05	0.74 ± 0.08
7.8	1.15 ± 0.15	0.73 ± 0.09	0.63 ± 0.11
Control	0.91 ± 0.06	0.72 ± 0.06	0.72 ± 0.06

DISCUSSION

From 92 fungi isolated from different ecosystems, the selection process performed in this paper indicated that *Leptosphaerulina* sp. is the fungal strain with the best decolorizing activity among the fungal strains isolated. At this moment, this is the first report demonstrating the decolorizing activity in the genre *Leptosphaerulina.* The ability of *Leptosphaerulina* sp. to decolorize solutions with dyes can be related to the presence of ligninolytic enzymes. Likewise, *L. chartarum* and *L. briosiana* have been described as ligninolytic enzymes producers (Simon et al. 1979; Sajben-Nagy et al. 2014).

The decolorizing activity observed by *Leptosphaerulina* sp. in remazol black was similar to *Bjerkandera adusta, Trametes versicolor, Funalia trogii* and *Irpex Lacteus,* strains recognized by their high decolorizing activity over Remazol Black (Mohorčič et al. 2006; Park et al. 2007; Máximo and Costa-Ferreira 2004). Similarly, *Leptosphaerulina* sp. decolorization percentage (>90 %) can be compared with the decolorization obtained by *P. simplicissimum* INCQS40211, *P. chrysosporium,* and *Ph. Tremellosa* over Turquoise Blue. In these fungal strains, the removal was above 94 % (Kirby et al. 2000; Bergsten-Torralba et al. 2009). Similarly, novacron red was decolorized by *Leptosphaerulina* sp. in a greater level than the reported by other fungal strains at similar or greater concentrations (≥200 ppm) of monoazoic dyes (Jarosz-Wilkołazka et al. 2002).

In terms of decolorization time, *Leptosphaerulina* sp. decolorized remazol black in times comparable with *Funalia trogii* (3 days); however, *F. Trogii* had a dye concentration (100 ppm) lower than *Leptosphaerulina* sp. (200 ppm) (Park et al. 2007). In other fungal strains, Remazol black decolorization was realized between 10 and 15 days and with a dye concentration lower than the used in this investigation (Mohorčič et al.2006; Park et al. 2007). Turquoise Blue decolorization exhibited a high decolorization time (3 days) (Fig. 3b). However, in this dye, the initial removal was performed by an adsorption process. The

Turquoise Blue high adsorption can be related to the copper molecule in the core of the dye structure. The adsorption of copper-related dyes has been observed in other fungal strains, as *P. simplicissimum* and *Penicillium oxalicum* (Bergsten-Torralba et al. 2009; Xin et al. 2010). In contrast, novacron red was decolorized in less days (9 days, 90 %); however, this period is considered as a fast decolorization for monoazoic dyes. Other fungal strains decolorized monoazoic dyes at the same concentration, during a time between 12–20 days (Swamy and Ramsay 1999; Toh et al. 2003). Faster decolorizations have been reported; nevertheless, these decolorizations used concentrations (80–100 ppm) less than this study (200 ppm) (Swamy and Ramsay 1999). Similar to turquoise blue, Novacron Red decolorization by *Leptosphaerulina* sp. exhibited two decolorization mechanisms, adsorption and enzymatic catalysis. Adsorption was the predominant mechanism during the 1st days; however, after the 3rd day, the principal mechanism was the enzymatic catalysis. The combination of these two removal mechanisms has also been described in *I. lacteus* and *Penicillium simplicissimum*in the decolorization of reactive red 198 and methyl red (Novotný et al. 2004; Bergsten-Torralba et al. 2009).

Leptosphaerulina sp. produced laccase and MnP to decolorize the three dyes, LiP was not observed in the cultures. It is important to remark that these enzymes were produced during all the culture; however, they reached their maximum production at different times. In remazol black, the largest enzyme production of laccase (650 U/L) and MnP(100 U/L) by *Leptosphaerulina* sp. was observed at day 15. Ligninolytic microorganisms that decolorize solutions with Remazol Black can be grouped in two sets, high laccase producers and high MnP producers.*Leptosphaerulina* sp. belongs to high laccase producers. The largest laccase activity exhibited by *Leptosphaerulina* sp. (650 U/L) is comparable with *T. versicolor* (577 U/L) and *C. dryhophyla* (335 U/L) (Baldrian and Šnajdr 2006). The MnP activity is also comparable with those species; however, the MnP activity in *Leptosphaerulina* sp. was small compared with other microorganisms, such as *F. trogii* (559 U/L) or *Geotrichum* sp. (300 U/L) (Máximo et al. 2003; Kariminiaae-Hamedaani et al. 2007). In turquoise blue experiment, the largest enzymatic production was observed at day 9 with a laccase activity of 512 U/L and a MnP activity of 100 U/L. The laccase activity dominance, in the treatment of phthalocyanine dyes, has been described in other fungal strains, such as *Gloeophyllum odoratum* (104 U/L), *Trametes pubescens* (391 U/L) and *F. trogii* (400 U/L) (Anastasi et al. 2010; Yesilada et al. 2010). The enzymatic production was observed during all the kinetic; nevertheless, the highest enzymatic activities of laccase and MnP were observed on day 13 with 647 U/L and 90 U/L, respectively. In the degradation of monoazoic dyes is not common to see a laccase production greater than MnP. Monoazoic dyes have been decolorized by fungal strains (*T. versicolor ATCC 20869, T. versicolor CNPR 8107*, and *I. lacteus*) with greater MnP activities than laccase (Novotný et al.2004; Toh et al. 2003).

The positive results obtained by *Leptosphaerulina* sp. and the differences obtained between the dyes are explained by the differences in the biochemistry of the decolorization process. The azoic dyes, Remazol black and Novacron red, are degraded principally by an enzymatic process (Michniewicz et al. 2008); whereas, phthalocyanine dyes (Turquoise Blue) are degraded by a mixture of biophysical and biochemical process. In this case, Turquoise Blue is initially absorbed by the cell wall and after that degraded by the ligninolytic enzymes (Swamy and Ramsay 1999; Kirby et al. 2000). Both types of degradation have a different physiological explanation. In the azoic dyes case, the chemical structure of azoic dyes is related with the presence of phenol groups. Phenol groups are the principal substrate for the ligninolytic enzymes; therefore, the enzymatic enzymes degrade the dye by using the structural similarity between azoic dyes and lignin (Kokol et al. 2007). On the other hand, phthalocyanine dyes are adsorbed in the membrane because they have a copper atom in their structure, copper atoms tend to be absorbed by ligninolytic fungus to further include them in the structure of their enzymes, in this case the dye molecule is absorbed and further degraded by the enzymatic enzymes (Jarosz-Wilkołazka et al. 2002).

The *Leptosphaerulina* sp. lyophilized-extracts (Fig. 4) transformed enzymatically all the dyes; however, this decolorization was dependent of the dye. *Leptosphaerulina* sp. enzymatic extracts achieved similar levels of remazol black decolorization than the commercial enzyme (80 %); however, this decolorization was less than the obtained by enzymatic extracts from *Pleurotus sajor-caju* and *Ganoderma lucidum*, using HBT as mediator (Murugesan et al. 2007). The enzymatic extracts transformed the remazol red into a reddish colour, this reddish coloration was described in the decolorization of Remazol Black by enzymatic extracts from *Pleurotus sajor-caju* and *Ganoderma lucidum* using N,N'-Bis-(1H-tetrazol-5-yl)-hydrazine (HBT) as mediator (Murugesan et al. 2007). In Turquoise Blue, the *Leptosphaerulina* sp. enzymatic extracts produced a decolorization of 56 % (Fig. 4). This enzymatic decolorization is similar to the obtained by extracts of *Dichomitus squalens* (60 %) over the copper (II) phthalocyanine dye (Svobodová et al. 2008). The low decolorization of monoazoic dyes by enzymatic extracts has been reported in acid yellow 49 and acid yellow 129 (Ciullini et al. 2008). In the future, the decolorization using *Leptosphaerulina* sp. enzymatic extracts can be improved by incrementing the MnP activity in the extracts and employing other types of chemical mediators, as HBT or vanillin.

Leptosphaerulina sp. displayed high decolorizing activity in the discharge and dyeing bath effluent (>90 %). This level of decolorization was achieved by the optimization of different factors. The principal ones were pH and effluent concentration. The importance of neutral or acid pH to realize the decolorization has been observed in *P. sajor-caju*, *B. adusta*, and *I. lacteus* (Mohorčič et al. 2006; Novotný et al. 2011). The dye concentration in the effluent was 50 times larger than the dye concentration tested in the previous experiments. *Leptosphaerulina* sp. was inhibited by the large dye concentration; however, at lower concentrations, *Leptosphaerulina* sp. decolorized efficiently the dyeing effluent. Effluent dilution reduces the growth and enzymatic production inhibitions (Mohorčič et al. 2006). Other factors such as nitrogen and glucose supplementation and initial inoculum also affected the decolorization process; however, they were less relevant than pH and effluent concentration. Glucose supplementation has been proved as an important factor in order to bio-remediate textile effluents with ligninolytic fungi (Novotný et al. 2011; Pakshirajan et al. 2011). In contrast, nitrogen supplementation had contradictory effects. *Aspergillus fumigatus* XC6 (Jin et al. 2007) and *Phanerochaete chrysosporium* (Pakshirajan et al. 2011) exhibited large decolorization, when nitrogen was supplemented; whereas, *C. versicolor* showed less decolorization after nitrogen supplementation (Asgher et al. 2009).

CONCLUSIONS

From 92 Colombian fungal isolates, *Leptosphaerulina* sp. exhibited the best decolorization percentage (>90 %) in solid and liquid cultures, and in agitated and un-agitated conditions. Additionally, *Leptosphaerulina* sp. decolorized two real effluents from textile industries (>90 %) under conditions of low pH and glucose/nitrogen supplementation. These decolorizations were associated with the production of significant quantities of laccase (650 U/L) and MnP (100 U/L). *Leptosphaerulina* sp. enzymatic extracts decolorized (20–80 %) the three dyes; however, they needed the presence of ABTS as mediator. The cytotoxicity test and the chromatographic techniques demonstrated the dyes' biochemical transformation and the absence of negative effects in the compounds produced by this biochemical process. *Leptosphaerulina*sp. is an interesting microorganism for future biotechnological applications, specially to treat waters from the textile industry.

DECLARATIONS

Authors' contributions

JP was responsible of experiment's conception and design, acquisition of data, analysis and interpretation of data, and manuscript drafting. XC was responsible of microbiological and molecular experimentation. SO was responsible of microbiological and enzymatic experiments. MY was responsible of analysis and interpretation of data and manuscript drafting and approval.

AM was responsible of final manuscript approval. All authors read and approved the final manuscript.

ACKNOWLEDGEMENTS

The authors acknowledge to The Research Direction of the National University of Colombia at Medellin (DIME) and the Biotechnology Postgraduate Program for the financial support to this project. Fabricato-Tejicondor S.A. for providing the dyes and effluents used in the study. The Mycotoxins, analysis and environmental processes, and molecular and cellular biology laboratories at the Faculty of Science at the National University of Colombia, at Medellin, for the technical and methodological support.

REFERENCES

1. *Abadulla E, Tzanov T, Costa S, Robra KH, Cavaco-Paulo A, Gubitz GM (2000) Decolorization and detoxification of textile dyes with a laccase from Trametes hirsuta. Appl Environ Microbiol 66:3357–3362*

2. *Anastasi A, Prigione V, Varese GC (2010) Industrial dye degradation and detoxification by basidiomycetes belonging to different eco-physiological groups. J Hazard Mater 177:260–267.*

3. *Angel TM (2002) Molecular biology and structure-function of lignin-degrading heme peroxidases. Enzyme Microb Technol 30:425–444.*

4. *Archibald FS, Bourbonnais R, Jurasek L, Paice MG, Reid ID (1997) Kraft pulp bleaching and delignification by Trametes versicolor. J Biotechnol 53:215–236.*

5. *Asgher M, Azim N, Bhatti HN (2009) Decolorization of practical textile industry effluents by white rot fungus Coriolus versicolor IBL-04. Biochem Eng J 47:61–65*

6. *Axelsson J, Nilsson U, Terrazas E, Aliaga TA, Welander U (2006) Decolorization of the textile dyes Reactive Red 2 and Reactive Blue 4 using Bjerkandera sp. Strain BOL 13 in a continuous rotating biological contactor reactor. Enzyme Microb Technol 39:32–37*

7. *Babot ED, Rico A, Rencoret J, Kalum L, Lund H, Romero J, del Río JC, Martínez ÁT, Gutiérrez A (2011) Towards industrially-feasible delignification and pitch removal by treating paper pulp with Myceliophthora thermophila laccase and a phenolic mediator. Bioresour Technol 102:6717–6722.*

8. *Baldrian P, Šnajdr J (2006) Production of ligninolytic enzymes by litter-decomposing fungi and their ability to decolorize synthetic dyes. Enzyme Microb Technol 39:1023–1029.*

9. *Bergsten-Torralba L, Nishikawa M, Baptista D, Magalhães D, Silva Md (2009) Decolorization of different textile dyes by* Penicillium *simplicissimum and toxicity evaluation after fungal treatment. Brazilian J Microbiol. 40:808–817*

10. *Chanagá Vera X, Plácido Escobar J, Marín Montoya M, Pérez Yepes, del Socorro María (2012) Native fungi with industrial dye degrading potential in the Aburrá valley, Colombia. Revista Facultad Nacional de Agronomía, Medellín. 65:6811–6821*

11. *Ciullini I, Tilli S, Scozzafava A, Briganti F (2008) Fungal laccase, cellobiose dehydrogenase, and chemical mediators: combined actions for the decolorization of different classes of textile dyes. Bioresour Technol 99:7003–7010.*

12. *Costa SM, Goncalves AR, Esposito E (2005) Ceriporiopsis subvermispota used in delignification of sugarcane bagasse prior to soda/anthraquinone pulping. In: Davison BH, Evans BR, Finkelstein M, McMillan JD (eds) Twenty-sixth symposium on biotechnology for fuels and chemicals. Humana Press, New York, pp 695–706*

13. *Couto S, Toca-Herrera J (2006) Lacasses in the textile industry. Biotechnol Mol Biol Rev 1:115–120*

14. *Dias A, Sampaio A, Bezerra R (2007) Environmental applications of fungal and plant systems: decolourisation of textile wastewater and related dyestuffs. In: Singh S, Tripathi R (eds) Environmental Bioremediation Technologies. Springer, Heidelberg, pp 445–463*

15. *dos Santos AB, Cervantes FJ, van Lier JB (2007) Review paper on current technologies for decolourisation of textile wastewaters: perspectives for anaerobic biotechnology. Bioresour Technol 98:2369–2385.*

16. *Erkurt EA, Ünyayar A, Kumbur H (2007) Decolorization of synthetic dyes by white rot fungi, involving laccase enzyme in the process. Process Biochem 42:1429–1435.*

17. *Gallego A, Torres F, Robledo S, Vélez ID, Carrillo L, Muñoz DL, Quiñones W, Fonnegra R, Roldán J, Valencia L (2006) ActividadLeishmanicida y tripanocida de Acacia farnesiana, Piper arieianum, P. subpedale, Sphagnum recurvum y vismia baccifera ssp. ferruginea. Actual Biol. 28:39–49*

18. *Hammel KE, Cullen D (2008) Role of fungal peroxidases in biological ligninolysis. Curr Opin Plant Biol 11:349–355.*

19. *Hao J, Song F, Huang F, Yang C, Zhang Z, Zheng Y, Tian X (2007) Production of laccase by a newly isolated deuteromycete fungusPestalotiopsis sp. and its decolorization of azo dye. J Ind Microbiol Biotechnol 34:233–240*

20. *Hernández-Luna C, Gutiérrez-Soto G, Salcedo-Martínez S (2008) Screening for decolorizing basidiomycetes in Mexico. World J Microbiol Biotechnol 24:465–473.*

21. *Jarosz-Wilkołazka A, Kochmańska-Rdest J, Malarcžyk E, Wardas W, Leonowicz A (2002) Fungi and their ability to decolourize azo and anthraquinonic dyes. Enzyme Microb Technol 30:566–572.*

22. *Jin X, Liu G, Xu Z, Tao W (2007) Decolorization of a dye industry effluent by Aspergillus fumigatus XC6. Appl Microbiol Biotechnol 74:239–243*

23. *Kariminiaae-Hamedaani H, Sakurai A, Sakakibara M (2007) Decolorization of synthetic dyes by a new manganese peroxidase-producing white rot fungus. Dyes Pigm 72:157–162.*

24. *Khindaria A, Yamazaki I, Aust SD (1996) Stabilization of the veratryl alcohol cation radical by lignin peroxidaseâe. Biochemistry 35:6418–6424.*

25. *Kirby N, Marchant R, McMullan G (2000) Decolourisation of synthetic textile dyes by Phlebia tremellosa. FEMS Microbiol Lett 188:93–96.*

26. *Kokol V, Doliška A, Eichlerová I, Baldrian P, Nerud F (2007) Decolorization of textile dyes by whole cultures of Ischnoderma resinosum and by purified laccase and Mn-peroxidase. Enzyme Microb Technol 40:1673–1677.*

27. *Mattinen M, Maijala P, Nousiainen P, Smeds A, Kontro J, Sipilä J, Tamminen T, Willför S, Viikari L (2011) Oxidation of lignans and lignin model compounds by laccase in aqueous solvent systems. J Mol Catal B Enzym 72:122–129.*

28. *Máximo C, Costa-Ferreira M (2004) Decolourisation of reactive textile dyes by Irpex lacteus and lignin modifying enzymes. Process Biochem 39:1475–1479.*

29. *Máximo C, Amorim MTP, Costa-Ferreira M (2003) Biotransformation of industrial reactive azo dyes by Geotrichum sp. CCMI 1019. Enzyme Microb Technol 32:145–151.*

30. *Michniewicz A, Ledakowicz S, Ullrich R, Hofrichter M (2008) Kinetics of the enzymatic decolorization of textile dyes by laccase fromCerrena unicolor. Dyes Pigm 77:295–302.*

31. *Mohorčič M, Teodorovič S, Golob V, Friedrich J (2006) Fungal and enzymatic decolourisation of artificial textile dye baths. Chemosphere 63:1709–1717.*

32. *Murugesan K, Dhamija A, Nam I, Kim Y, Chang Y (2007) Decolourization of Reactive Black 5 by laccase: optimization by response surface methodology. Dyes Pigm 75:176–184*

33. *Novotný Č, Svobodová K, Kasinath A, Erbanová P (2004) Biodegradation of synthetic dyes by Irpex lacteus under various growth conditions. Int Biodeterior Biodegrad. 54:215–223.*

34. *Novotný Č, Svobodová K, Benada O, Kofroňová O, Heissenberger A, Fuchs W (2011) Potential of combined fungal and bacterial treatment for color removal in textile wastewater. Bioresour Technol 102:879–888*

35. *Nyanhongo GS, Gomes J, Gübitz GM, Zvauya R, Read J, Steiner W (2002) Decolorization of textile dyes by laccases from a newly isolated strain of Trametes modesta. Water Res 36:1449–1456.*

36. *Pakshirajan K, Sivasankar A, Sahoo NK (2011) Decolourization of synthetic wastewater containing azo dyes by immobilized Phanerochaete chrysosporium in a continuously operated RBC reactor. Appl Microbiol Biotechnol 89:1223–1232*

37. *Park C, Lee M, Lee B, Kim S, Chase HA, Lee J, Kim S (2007) Biodegradation and biosorption for decolorization of synthetic dyes byFunalia trogii. Biochem Eng J 36:59–65.*

38. *Risna RA, Suhirman S (2002) Ligninolytic enzyme production by Polyporaceae from Lombok Indonesia. Fungal Divers 9:123–134*

39. *Robinson T, Chandran B, Nigam P (2001) Studies on the production of enzymes by white-rot fungi for the decolourisation of textile dyes. Enzyme Microb Technol 29:575–579.*

40. *Sajben-Nagy E, Manczinger L, Škrbić B, Živančev J, Antić I, Krisch J, Vágvölgyi C (2014) Characterization of an extracellular laccase ofLeptosphaerulina chartarum. World J Microbiol Biotechnol 30:2449–2458*

41. *Salony S, Mishra S, Bisaria V (2006) Production and characterization of laccase from Cyathus bulleri and its use in decolourization of recalcitrant textile dyes. Appl Microbiol Biotechnol 71:646–653.*

42. *Sergio R (2006) Laccases: blue enzymes for green chemistry. Trends Biotechnol 24:219–226*

43. *Simon LT, Bishop DS, Hooper GR (1979) Ultrastructure and cytochemical localization of laccase in two strains of Leptosphaerulina briosiana (Pollaci) Graham and Luttrell. J Bacteriol 137:537–544*

44. *Svobodová K, Majcherczyk A, Novotný Č, Kües U (2008) Implication of mycelium-associated laccase from Irpex lacteus in the decolorization of synthetic dyes. Bioresour Technol 99:463–471.*

45. *Swamy J, Ramsay JA (1999) The evaluation of white rot fungi in the decoloration of textile dyes. Enzyme Microb Technol 24:130–137*

46. *Telke AA, Kalyani DC, Dawkar VV, Govindwar SP (2009) Influence of organic and inorganic compounds on oxidoreductive decolorization of sulfonated azo dye CI Reactive Orange 16. J Hazard Mater 172:298–309*

47. *Toh Y, Yen JJL, Obbard JP, Ting Y (2003) Decolourisation of azo dyes by white-rot fungi (WRF) isolated in Singapore. Enzyme Microb Technol 33:569–575.*

48. *Weisburger JH (2002) Comments on the history and importance of aromatic and heterocyclic amines in public health. Mutat Res 506–507:9–20.*

49. *Xin B, Chen G, Zheng W (2010) Bioaccumulation of Cu-complex reactive dye by growing pellets of Penicillium oxalicum and its mechanism. Water Res 44:3565–3572*

50. *Yesilada O, Yildirim SC, Birhanli E, Apohan E, Asma D, Kuru F (2010) The evaluation of pre-grown mycelial pellets in decolorization of textile dyes during repeated batch process. World J Microbiol Biotechnol 26:33–39*

51. *Zouari-Mechichi H, Mechichi T, Dhouib A, Sayadi S, Martínez AT, Martínez MJ (2006) Laccase purification and characterization fromTrametes trogii isolated in Tunisia: decolorization of textile dyes by the purified enzyme. Enzyme Microb Technol 39:141–148.*

Chapter 10

Effects of a Textile AZO Dye On Mortality, Regeneration, and Reproductive Performance of the Planarian, *Girardia Tigrina*

Alyson Rogério Ribeiro and Gisela de Aragão Umbuzeiro*

School of Technology, State University of Campinas - UNICAMP, Paschoal Marmo Street 1888, Limeira, SP 13484-332, Brazil

ABSTRACT

Background

Many dyes can be considered emerging contaminants. The most widely used dyes belong to the class of azo compounds, some of which are known to have toxic and genotoxic properties. They are used in great quantities in textile activities and are of environmental concern because of their potential discharge in water. Planarians have been successfully used as test organisms in hazard evaluation of different chemicals, and we demonstrate the suitability of *Girardia tigrina* for laboratory testing. The aim of this work was to evaluate the suitability of the planarian species *G. tigrina* to assess the ability of the azo textile dye disperse red 1 to cause acute toxicity and adverse effects in the regeneration and reproduction of newborn and adult specimens.

Results

Disperse red 1 presented a median LC_{50} of 75 and 152 mg/L, respectively, for newborns and adults of *G. tigrina*, showing that newborns are twice as susceptible to the dye. Uncoordinated movements, irregular twists, colored skin, increased mucous production, and regenerative delays were observed after dye exposure at sub-lethal concentrations.

Conclusions

A no-observed-adverse-effect concentration (NOAEC) of 0.1 mg/L could be determined for disperse red 1 based on the fecundity test. Zinc seems to be a suitable positive control for monitoring the sensitivity in *G. tigrina* tests within only 24 h of exposure. This study demonstrates the applicability of *G. tigrina* tests in the hazard evaluation of water contaminants, such as azo dyes.

Keywords: Regeneration Planarians Reproductive performance Azo dye Disperse red 1 *Girardia tigrina*

BACKGROUND

Synthetic dyes can be considered emerging contaminants because they are potentially toxic and have been found in the aquatic environment and there are no regulations stating the maximum allowable concentrations in water to ensure the protection of aquatic biota and human health. The world production of key dyes is estimated to be more than 10 tons per year [1]. The most widely used dyes belong to the class of azo compounds derived from certain aromatic amines and are used in products such as textiles, foodstuffs, cosmetics, house products, paints, and inks. This is significant because some of these dyes are known to have potential toxic and genotoxic properties [2,3]. The use of azo dyes to the color of textiles is of considerable interest, as this can raise environmental concerns because of the high volume of water involved in the dyeing process. When a dye is used in this process, a portion does not undergo bonding to the fibers thus remaining in the water bath [4]. As a consequence, high volumes of wastewaters containing dyes and related auxiliaries are produced and may be released into the environment. For instance, for 10 kg of polyester fabric, 100 g of disperse red 1 will be dissolved in 200 L of water, of which are round 1% of the dye will remain in the water bath at the end of the process. Hence, around 200 L of wastewater containing 1 g of the dye will be generated. It is also known that conventional treatments involving aerobic lagoons or activated sludge are not efficient in the removal or biological degradation of these dyes [5] and therefore, the dyes are still present in the final effluent or in the sludge of the treatment plant [6,7]. Disperse dyes used to dye synthetic fibers are generally sparingly water-soluble compounds, but they can be found in the water column, because of their commercial formulation that includes surfactants needed for the dyeing process. Some disperse dyes have been found in the aquatic environment [8], and their presence was related to the observed mutagenic activity of the water and sediments [6,7,9].

Recently, Ferraz et al. [10] showed that the textile dye disperse red 1 ((*N*-ethyl-*N*-(2-hydroxyethyl)-4-(4-nitrophenylazo) aniline) is highly toxic to aquatic invertebrates, in addition to being mutagenic. Vacchi et al. [11] studied a commercial disperse red 1 dye product and found that the ecotoxicity of the commercial preparation was similar to the dye itself. The median effective concentration (EC_{50}) for *Daphnia* was 0.1 mg/L; therefore, this dye can be classified as highly toxic to aquatic organisms according to the Globally Harmonized System of Classification and Labelling of Chemicals (GHS) [12]. More recently, studies have been performed to estimate the concentrations of dyes in surface waters [13], Zocollo et al.'s submitted article. The authors determined the concentrations of disperse azo dyes in river and effluent samples collected in Brazil and showed that they were present in the levels of nanograms to micrograms per liter. However, more studies are required to estimate the concentration levels in the aquatic environment, especially in countries where dyeing activities are more intense, like in India, China, and Brazil.

Planarians have been used as test organisms in the hazard evaluation of different chemicals using different endpoints such as mortality, regeneration, micronucleus (MN) frequency, and enzymatic activity [14-21]. Planarians generally reproduce by transverse fission, but some species, like *Girardia tigrina* [22], also have hermaphroditic sexual organs and can generate cocoons of fertilized eggs [23]. Therefore, reproduction impairment can also be used as an endpoint in ecotoxicity tests with planarians [20,22].

Because of its ability to regenerate [24,25], planarians have been used to verify if chemicals can interfere with this process [17,18,26-30]. A small body fragment can generate an intact planarian due the presence of totipotent stem cells called neoblasts, which migrate from the parenchymal tissue to the injured site and differentiate by mitosis to other planarian cell types [31,32].

There have been a number of publications that examined the response of planarians to chemical exposures using molecular and enzymatic approaches [21,33-35]. We focused our study towards three different low-cost and environmentally relevant endpoints: mortality, regeneration, and reproduction. The aim of this work was to evaluate the suitability of the planarian species *G. tigrina* as a test organism to assess the acute toxicity of the azo textile dye, disperse red 1, to newborns and adults and to verify its ability to cause adverse effects on regeneration and cocoon production in the exposed animals. Disperse red 1 was selected because it has been found in river waters that receive textile effluents in the region of Americana, São Paulo, Brazil (Zocolo et al.'s submitted article).

RESULTS AND DISCUSSION

G. tigrina showed an acceptable sensitivity to $ZnSO_4$, with a 24-h median lethal concentration of LC_{50} of 1.6 ± 0.2 mg/L expressed in Zn^{2+} and a variation coefficient of 12% in seven replicate acute newborn toxicity tests (Table 1). Chromium salts have previously been used to monitor sensitivity of laboratorial culture [15,36,37]. We also used chromium for this purpose but zinc provided better repeatability and faster responses (24 h) (Table 1). Although the mean LC_{50} values after 48 and 96 h for chromium are similar to those reported by Preza and Smith [15], the coefficient of variation (66%) in our study was higher than obtained by those authors (22%). This higher variation could be explained by the higher instability of chromium in solutions compared to zinc or by response differences between the planarian populations.

Table 1. Mean lethal concentrations (50%) (LC_{50}) of chromium and zinc in newborn ***G. tigrina*** acute tests for sensitivity assessment

	LC_{50} (mg/L)							
Test	Cr^{6+}				Zn^{2+}			
	24 h	48 h	72 h	96 h	24 h	48 h	72 h	96 h
I	13	11	8	7	2	2	2	2
II	11	11	11	7	1.6	1.6	1.6	1.6
III	*	27	20	11	1.6	1.6	1.6	1.6
IV	*	29	18	15	1.6	1.6	1.6	1.6
V	18	8	5	3	1.3	1.3	1.3	1.3
VI	15	10	5	4	1.6	1.6	1.6	1.6
VII	14	5	2	5	1.7	1.7	1.7	1.7
Mean	14.2	14.4	9.8	7	1.6	1.6	1.6	1.6
SD[a]	2.5	9.5	6.8	4.6	0.2	0.2	0.2	0.2
CV (%)[b]	18	66	69	66	12	12	12	12

Asterisk indicates LC_{50}, not calculable for the statistical method employed. [a]SD standard deviation of the mean. [b]Coefficient of variation. The tests indicated as I, II, III, IV, V, VI, and VII correspond to independent experiments.

The LC_{50} values for disperse red 1 dye obtained with newborns and adults are presented in Table 2, and as expected, toxicity increased with exposure time. Three independent tests were performed, and the coefficients of variation of the means were below 17%, which was considered acceptable for ecotoxicological assays according to the recommendation of Environment Canada [38]. No mortality was observed when adults were exposed during the first 24 h but several animals showed uncoordinated movements and increased mucus production at all tested concentrations. Mean 96-h LC_{50} values of 75 ± 7.2 mg/L ($n = 3$) and 152 ± 5.8 ($n = 3$) mg/L were obtained for newborns and adults, respectively (Table 2). These results confirmed the findings of Preza and Smith [15] which showed that newborns were more sensitive than adults. Knakievicz and Ferreira [20] attributed this age response to different body surface/volume ratios of newborn organisms.

Table 2. Mean lethal concentrations (50%) (LC_{50}) of commercial disperse red 1 in newborn and adult ***G. tigrina*** tests

	LC_{50} (mg/L)							
Test	**Newborn**				**Adult**			
	24 h	**48 h**	**72 h**	**96 h**	**24 h**	**48 h**	**72 h**	**96 h**
I	127	127	74	67	*	201	179	159
II	120	91	79	79	*	207	156	148
III	131	111	87	80	*	198	162	150
Mean	126	110	80	75		202	167	152
SD[a]	5.6	18	6.5	7.2		4.6	12	5.8
CV (%)[b]	4.4	16.4	8.1	9.6		2.2	7.2	3.8

Asterisk indicates no mortality was observed. [a]SD standard deviation of the mean. [b]Coefficient of variation. The tests indicate as I, II, and III correspond to independent assays.

Ecotoxicological data of azo textile dyes are scarce in the literature. Ferraz et al. [10] tested disperse red 1 (95% purity) and obtained an $EC_{50}$of 0.13 mg/L in a *Daphnia similis* acute toxicity test. Recently, the same commercial product that was analyzed in this study was used in acute toxicity tests with *D. similis* and *Hydra attenuata* [11]. The authors showed that the surfactant and other impurities present in the commercial dye did not influence the observed

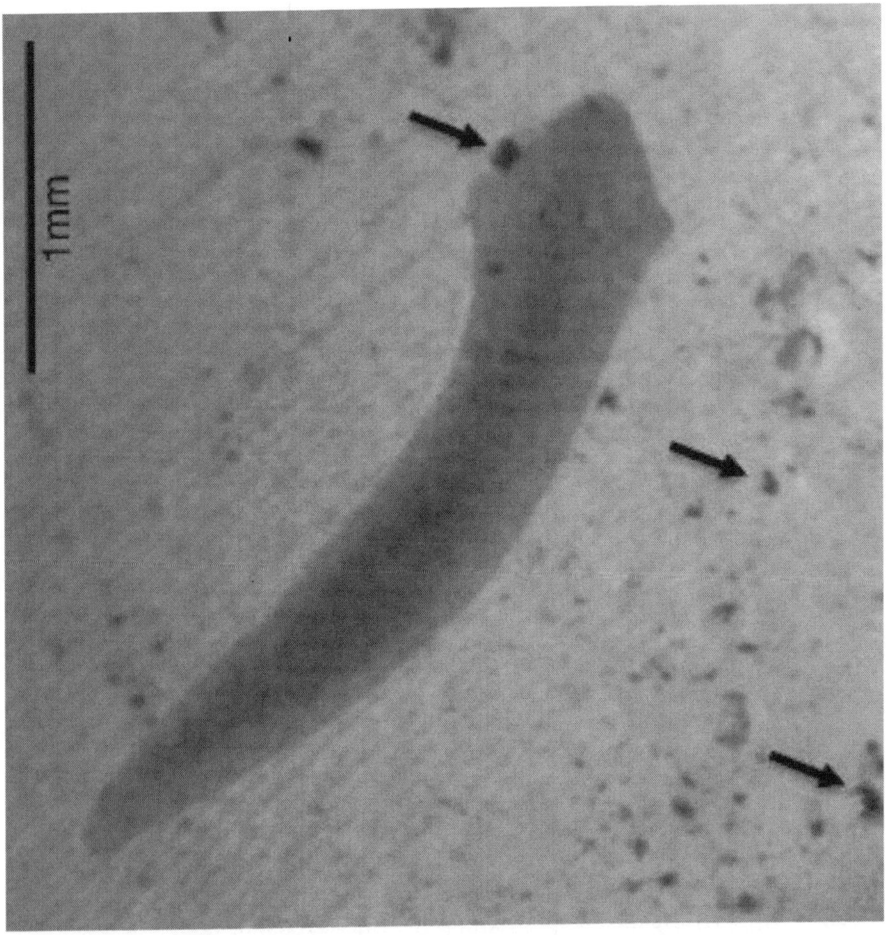

Figure 1. Picture of a newborn of *G. tigrina* after 96-h exposure to 10 mg/L of commercial disperse red 1 dye . Red-colored skin can be observed. Arrows indicate the precipitation and agglomeration of the dye. The image is 50 times increased.

toxicity, at least for *Daphnia*. The observed effects were related to the main dye, which represents 60% of the commercial product. In that study, it was not possible to verify if the acute toxicity observed for Hydra was related also to the main dye, but Hydra was 15 times less sensitive than Daphnia to the commercial dye [11].

Aggregation and precipitation of the dye were observed at concentrations higher than 50 mg/L after the first 24 h of exposure. This is expected because the commercial dye contains surfactants. Therefore, when the concentration of the disperse dye increases, they can form aggregates and settle on the bottom. This behavior is also expected to occur in the aquatic systems [39]. Because planarians slide on the surface of the testing containers [40], the animals were exposed to the precipitated/agglomerated dye in addition to the dye remaining in the solution. This natural surface-contact behavior of the planarians is an important consideration when assessing the toxicity of compounds that are sparingly soluble in water.

After 24 h of exposure in the acute toxicity tests, newborns (Figure 1) and adults (not shown) exposed to concentrations higher than 10 mg/L showed red-colored skin, especially at the encephalic region. This could be explained by dermal or cilia dye adsorption during locomotion and/or skin respiration. The exposed organisms showed uncoordinated movements, irregular twists and increased mucous production after the first 24-h exposure. The mucous resulted in dye precipitation and aggregation around the animal bodies, forming what appeared to be a body capsule, which was abandoned when the planarians started to move. Planarian behavioral effects, such as increased mucus production, irregular twisting, and body contractions, have previously been associated with animal responses to toxicants [26,41,42].

In the regeneration assay, the first changes were observed after 24 h at the highest concentration tested, 200 mg/L. After 120 h, 95% of control planarians fully regenerated (see Figure 2e). The no-observed-adverse-effect concentration (NOAEC) for the dye was 10 mg/L and the lowest-observed-adverse-effect concentration (LOAEC) was 50 mg/L (Table 2). The regenerated planarians showed also red-colored skin, behavioral changes, and increased mucous production at concentrations higher than 10 mg/L. Regenerative delay started to occur at 50 mg/L, while at the highest concentration (200 mg/L), eight of the 20 organisms died within 96 h of exposure, and the ones that survived presented severe regenerative delay after 120 h of exposure (Table 3) with parenchymal loss due to lack of wound cicatrization (see Figure 2c,d). When cicatrization occurred, some delays in posterior regenerative stages were noticed. In seven animals the auricle development was inhibited or delayed (see Figure 2a and 2b). For successful regeneration of the lost body parts, planarians need an intact nervous system to control the migration of the normal neoblasts [43,44] that may have been affected by the dye. Also, the mutagenic properties of the dye [10,11] could affect the normal mitotic process required for the cell replication, which is needed for regeneration. The planarian regeneration assays are useful for analyzing the effects of chemicals on cell differentiation and molecular organization [45]. Thus the migration and proliferation of neoblasts as well as other steps required to *G. tigrina* regeneration may have been affected by the azo dye.

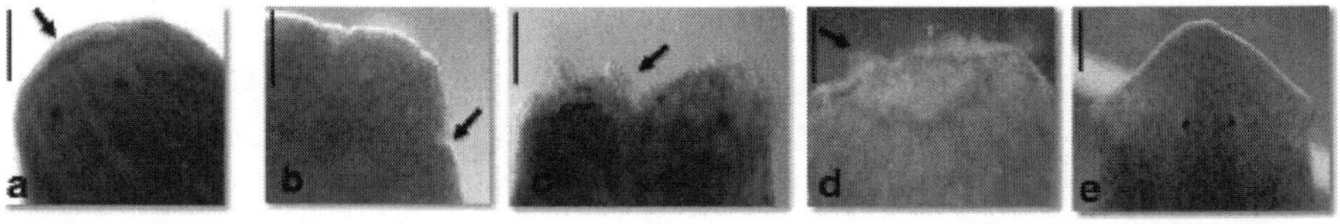

Figure 2. Morphological effects of different concentrations of commercial disperse red 1 dye in *G. tigrina* . (a) Individual without full body cicatrization, showing eyespots and no auricles (phase z) - after 120-h exposure at 200 mg/L. **(b)** Late development of the auricle (phase d) after 120 h at 50 mg/L. **(c, d)** Individuals with delayed and abnormal cicatrization (phase t) - after 120 h at 200 mg/L. **(e)** A fully developed planarian with no effect (negative control). Arrows are guides to the eye and the bar on the left of each figure represent 1 mm.

Table 3. Regenerative results for ***G. tigrina*** exposed to different concentrations of the commercial disperse red 1

Exposure time (h)	Planarians/regeneration steps (mg/L)					
	Control	10	50	100	150	200
12	20[a] (a)[b]	20(a)	20(a)	20(a)	20(a)	20(a)
24	20(b)	20(b)	20(b)	20(b)	20(b)	3(t) 17(b)
48	20(c)	20(c)	20(c)	20(c)	20(c)	3(t) 17(c)
72	20(de)	20(de)	20(de)	2(c) 18(de)	2(c) 18(de)	3(t) 17(de)
96	20(f)	20(f)	3(d) 17(f)	5(d) 15(f)	2(d) 18(f)	3(t) 9(d) 8(f)
120	1(f) 19(g)	20(g)	2(f) 18(g)	5(f) 15(g)	5(f) 15 (g)	8 dead 3(t) 3(z) 4(d) 2(g)
Pd^{c}_{120h}		−1	1	4	4	9

[a]Number of animals. [b]Regeneration steps: (a) decapitated bodies, (b) beginning of blastema formation, (c) regular blastema, (d) beginning of auricle formation, (e) beginning of eyespots formation, (f) auricle and eyespots clearly defined, (g) full head formation, (t) body section without cicatrization, (z) without body cicatrization, with eyespots, and without auricle. [c]Pd is the number of planarians with delayed regeneration subtracted from the number of delayed planarians in the control, Pd greater or equal to 1 is considered as adverse, for more information see 'Methods' section.

Freshwater planarian reproduction is a complex process and some details are still unknown [22]. These animals have a many-sided reproductive organ with ovaries and hormones. Animals can be fertilized after mating [23,46]. Eggs, formed after fertilization, as well as vitelline cells, are enveloped by an oval wall called cocoons [47]. Hatching occurs after different incubation periods depending on the species and temperature [40,48]. The time between crosses and cocoon lay is still unknown. In our experiments, the fecundity index recorded for the animals exposed to 1 mg/L of disperse red 1 (stage 2, exposure) was statistically significantly lower than the control (Figure 3). Also, after the first week of exposure, animals from the treated group G4 (1 mg/L) presented colored skin and alterations in behavior during feeding, such as continuous exposure of the pharynx and limited mobility. The pharynx is only exposed during a planarian's feeding process, and after feeding this organ returns to planarian's gastric cavity [40]. However, we observed that planarians, when exposed to the dye, showed retractile pharynx movements even when they were far from the food. Planarian reproductive success is related to food availability, demographic density, and temperature [22,48,49]. During the entire exposure time, the animals from G4 group were not able to eat and the cocoons produced during that time may have been generated using their remaining energy [36,50]

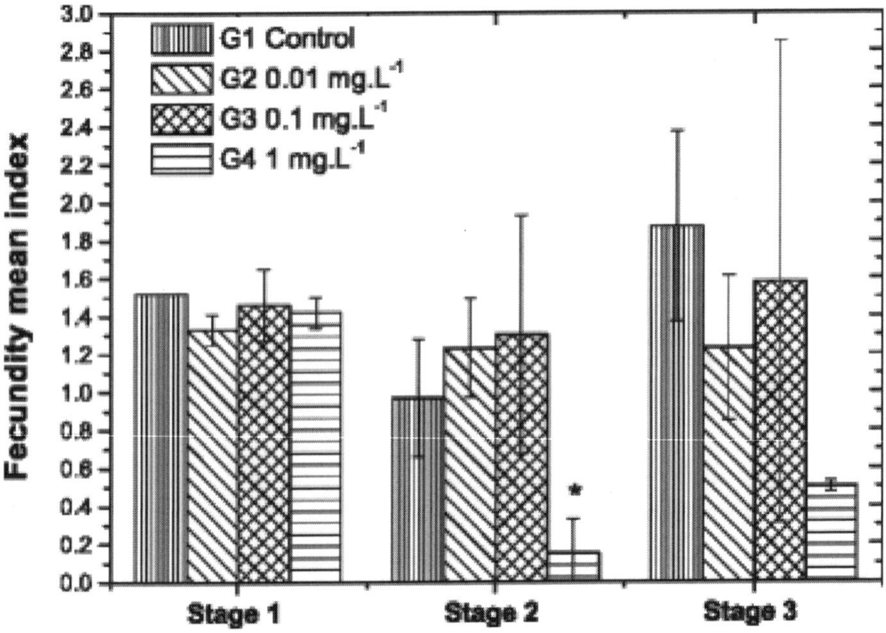

Figure 3. Mean fecundity index of *G. tigrina* planarian exposed to commercial textile dye disperse red 1 (control and exposed groups). At stage 1, exposure system adaptation, organisms were monitored for 2 weeks in maintenance medium. At stage 2, organisms were exposed for 5 weeks. At stage 3, recovery, exposed organisms were monitored for 2 weeks more in maintenance water. Bars represent the standard deviation of the mean. *Significant at $p \leq 0.05$.

During stages 1 (before exposure) and 3 (recovery following exposure) no statistical differences were observed in the fecundity indexes among the groups, although fecundity capacity did not seem to be fully recovered when animals were exposed to 1 mg/L during stage 3 (Figure 3). The addition of the recovery stage was also useful to analyze if the planarians were able to remove the impregnated or absorbed dye from their bodies and if the effect on the pharynx's retractability was permanent. After 2 weeks of recovering in maintenance water more than 80% of the exposed animals showed normal skin color as well as normal pharynx function and feeding behavior. Therefore, disperse red 1 interferes with *G. tigrina*'s mobility and capacity for food identification at 1 mg/L exposure concentration. In sum, disperse red 1 dye adversely affects survival, fecundity, regeneration, and feeding behavior in planarians. Based on the most critical endpoint (fecundity index) it is possible to suggest a no-observed-adverse-effect-concentration (NOAEC) of 0.1 mg/L for disperse red 1. As observed in other studies [15,17,21,22,49], the viability of *G. tigrina* culture under low-cost laboratorial conditions was confirmed. Our results also show the repeatability of the applied protocol at least for zinc and disperse red 1.

Planarians can be considered an interesting model for ecotoxicological assessment of environmental contaminants because it allows the assessment of different sub-lethal effects, such as increase in mucus production and changes in behavior besides the evaluation of adverse effects on the regeneration process and reproduction as showed in this work.

CONCLUSIONS

Disperse red 1 presented a median LC_{50} of 75 and 152 mg/L, respectively, for newborns and adults of *G. tigrina*, showing that newborns are more susceptible to the dye than adults. Uncoordinated movements, irregular twists, colored skin, and increased mucous production were observed after dye exposure at sub-lethal concentrations. A no-observed-adverse-effect concentration (NOAEC) of 0.1 mg/L of disperse red 1 could be determined based on the fecundity test. Zinc seems to be a suitable reference substance for monitoring the sensitivity in *G. tigrina*tests within only 24 h of exposure. This study successfully demonstrates the applicability of *G. tigrina* tests in the hazard evaluation of azo dyes, a relevant example of an emerging contaminant. The results obtained in this work may also be used for the determination of safe environmental concentrations of disperse red 1 for the aquatic system.

METHODS

Chemical

The powdered commercial textile dye product disperse red 1 dye (*N*-Ethyl-*N*-(2-hydroxyethyl)-4-(4-nitrophenylazo) aniline; CAS No 2872-52-8), was purchased from PCIL© Dyes, São Paulo, Brazil. The product was chemically characterized by Vacchi et al. [11] and contains 60% of the main dye disperse red 1, 20% of other azo dye products, and 20% of a surfactant. Zinc sulfate ($ZnSO_4.7H_2O$; CAS

No 7440-20-0; CAQ Casa de Química©, Diadema, São Paulo, Brazil) and potassium dichromate ($K_2Cr_2O_7$; CAS No 7778-50-9; Cromoline Química Fina©, Diadema, São Paulo, Brazil) in analytical grade were used as reference substances to monitor the sensitivity of the planarian culture. Dye and metal testing solutions were prepared in planarian maintenance water (see description below) in concentrations determined by preliminary tests.

Animals

Specimens of *G. tigrina* collected in a pristine pond in Rio Claro, São Paulo, Brazil [17], have been cultivated in the Ecotoxicology and Environmental Microbiology Laboratory (LEAL) in the University of Campinas, Limeira, Brazil, since 2005. Descendant animals were cultured by sexual reproduction at room temperature (22 to 25°C) under a light/dark photoperiod of 16:8 h, and fed once a week with bovine liver. The water for culturing the planarians and conducting experiments was designated as maintenance water and was prepared with dechlorinated tap water and 45 mg $CaCO_3$/L, pH 7.6 ± 0.3, and with gentle aeration [37] using an aquarium pump. Water was completely renewed each week after feeding.

Toxicity tests

Three environmentally relevant endpoints were investigated in this study through different toxicity tests: acute toxicity (mortality), regeneration assay and reproductive performance. Acute toxicity tests were carried out with adult and newborn planarians. Animals of more than 3 months of age and visible cocoon production capacity were considered adults, and animals with a maximum age of 10 days were considered newborns [15]. Each independent newborn toxicity test was carried out in duplicate using seven animals placed in 50 mL of test solution for 96 h [15]. In the adult toxicity tests healthy animals were randomly selected and five planarians were exposed to 100 mL of test solution. Each independent experiment was performed in triplicate [37]. In both the newborn and adult tests, the test solution was not renewed and the animals were not fed. Negative controls with maintenance water and positive controls with $ZnSO_4$ and $K_2Cr_2O_7$ were also performed. The nominal concentration used in the positive controls using chrome were 0.9, 1.8, 3.5, 7, 10, 18 and 35 mg/L expressed as Cr^{6+} and using zinc were 0.1, 0.2, 0.5, 1.1, 2.2, 4.5, and 6.8 mg/L, expressed as Zn^{2+}. Mortality, body degeneration, change in behavior and morphological alterations were monitored every 24 hours under a stereo-microscope (Stemi 2000-C, Zeiss©, Oberkochen, Germany). The nominal dye concentrations tested were 10, 50, 100, 150, and 200 mg/L for newborns and 50, 100, 150, 200, and 250 mg/L for adults.

For the regeneration assay, the heads of adult planarians were removed in the region behind the auricles using blades, and the decapitated animals were used immediately [14]. The animals were fed 1 day before the test started to ensure that they would have sufficient nutrients to regenerate. This is important because *G. tigrina* has the ability to store vital nutrients [36] and the digestion and assimilation of food, in particular liver, takes about 197 h [50]. The concentrations of disperse red 1 used (10, 50, 100, 150 and 200 mg/L) were selected based on the toxicity test results. In each independent experiment, each concentration was tested using four replicates; five decapitated animals per replicate were put in 20 mL of solution. Each head regeneration step [17,31] (designated, a decapitated bodies; b beginning of blastema formation; c regular blastema; d beginning of auricle formation; e beginning of eyespot formation; f auricle and eyespots clearly defined; g full head formation; and t body section without cicatrization, z without body cicatrization, with eyespots, and without auricle) were observed under the stereoscope daily, followed by renewal of the dye test solution [20]. Regeneration of decapitated planarians in maintenance water was used as a negative control. Animals were not fed, and they were kept in cultivation conditions for 120 h for full regeneration of the encephalic region [17].

The reproductive performance of adult specimens was evaluated based on a method described elsewhere [20]. The single experiment included a preexposure stage for adaptation purposes. We also followed the recovery capacity in a postexposure period, without the presence of the toxicant. Twenty-five healthy adult animals were randomly placed into plastic vessels containing 1 L of each test solution, without replicates. The test was carried out in three stages: at stage 1 (adaptation), all organisms were exposed only to maintenance medium and were monitored for 2 weeks. At stage 2 (toxicant exposure), three groups of organisms were exposed to sub-lethal concentrations (0.01, 0.1, or 1 mg/L) of the commercial product. We also included a negative control, without the dye. Concentrations were selected based on the acute toxicity test results. This stage lasted five weeks. At stage 3 (recovery), chemical exposure was interrupted at the end of the 5 weeks; animals were transferred into maintenance water and monitored for an additional 2 weeks. Observations on cocoon production, animal behavior, and mortality were made during all stages. Identical cultivation maintenance procedures were used for all groups, with weekly feeding followed by total test solution renewal, and the counting and segregation of cocoons.

During the toxicity, reproductive, and regeneration experiments, any changes in animal behavior, skin color, and mucous production were registered on a daily basis. The retractability of the pharynx during feeding was also monitored.

Statistical analysis and expression of the results

For acute toxicity tests, the trimmed Spearman-Karber method [51] was used to calculate the LC_{50}, with 95% confidence interval, for every 24 h of exposure, until 96 h. In the regeneration tests, a 120-h delayed regeneration value (Pd120h) was calculated for each concentration. To obtain this value, the number of delayed regeneration planarians in the control was subtracted from the treated number. The NOAEC was the higher concentration that presented a Pd120h <1, and the LOAEC was the lowest concentration that presented a Pd120h ≥1.

In the reproductive performance assay, the mean fecundity index for each group in each stage was calculated using a method adapted from Knakievicz et al. [22]. Mean fecundity indexes were calculated by the total number of cocoons divided by the number of live planarians in every analyzed week, divided by the total experimental time, in weeks, at the end of each step. To compare the different treatments, analysis of variance (ANOVA) followed by the Tukey test was performed using the Origin® Pro8 program. The significance level was $p \leq 0.05$.

DECLARATIONS

Acknowledgements

This work was part of the Thematic Project no. 2008/10449-7 financed by FAPESP - Fundação de Amparo a Pesquisa do Estado de São Paulo. ARR thanks FAPESP also for the master fellowship, process no. 2009/12737-2. ARR and GAU thank Alain Devaux, Sylvie Bony, Maria Beatriz Bohrer-Morel, Errol Zeiger, and Harold Freeman for the valuable suggestions, Francine Inforçato Vacchi for helping with the manuscript organization, and Erika Rabello Moretti for helping with the reproductive assay.

Competing interests

The authors declare that they have no competing interests.

Authors' contributions

ARR performed all experiments and statistical analysis and drafted the manuscript. GAU participated in the experimental design, supervised the research work, and was responsible for the final elaboration of the final manuscript. GAU also provided technical and financial support for this study. Both authors read and approved the final manuscript.

REFERENCES

1. *ETAD: The Ecological and Toxicological Association of Dyes and Organic Pigments Manufacturers, Annual Review. 2012. Issued April 2013 [*www.etad.com/documents/Downloads/publications/etad_+ar+2012_final_web.pdf*]*

2. *Novotný C, Dias N, Kapanen A, Malachová K, Vándrovcová M, Itävaara M, Lima N:* **Comparative use of bacterial, algal and protozoan test to study toxicity of azo- and anthraquinone dyes.** *Chemosphere 2006,* **63:***1436–1442.*

3. *Chung KT, Cerniglia CE:* **Mutagenicity of azo dyes: structure-activity relationships.** *Mutat Res-Rev Genet 1992,* **277***(Suppl 3):201–220.*

4. *Hanger K: Industrial dyes: Chemistry, Properties and Applications, Health and Safety Aspects. Germany: Wiley-VCH; 2003.*

5. *USEPA: Aerobic and Aerobic Treatment of C.I. Disperse Blue 79. US Department of Commerce. National Technical Information Service (NTIS) vols I and II. EPA/600/2-89/051 (PB 90–111642); 1989.* http://nepis.epa.gov/Exe/ZyPURL.cgi?Dockey=2000TM97.txt.

6. *Umbuzeiro GA, Roubicek DA, Rech CM, Sato MIZ, Claxton LD:* **Investigating the sources of the mutagenic activity found in a river using the Salmonella assay and different water extraction procedures.** *Chemosphere 2004,* **54***(Suppl 11):1589–1597.*

7. *Umbuzeiro GA, Freeman HS, Warren SH, Oliveira DP, Terao Y, Watanabe T, Claxton LD:* **The contribution of azo dyes to the mutagenic activity of Cristais River.** *Chemosphere 2005,* **60***(Suppl1):55–64.*

8. *Maguire RJ:* **Occurrence and persistence of dyes in a Canadian River.** *Water Sci Technol 1992,* **25:***264–270.*

9. *Oliveira DP, Carneiro PA, Rech CM, Zanoni MV, Clauton LD, Umbuzeiro GA:* **Mutagenic compounds generated from the chlorination of disperse azo dyes and their presence in drinking water.** *Environ Sci Technol 2006,* **40***(Suppl 21):6682–6689.*

10. *Ferraz ERA, Umbuzeiro GA, De Almeida G, Caloto Oliveira A, Chequer FMD, Zanoni MVB, Dorta DJ, Oliveira DP:* **Differential toxicity of disperse red 1 and disperse red 13 in the Ames test, HepG2 cytotoxicity assay, and Daphnia acute toxicity test.** *Environ Toxicol2011,* **26:***489–497.*

11. *Vacchi FI, Albuquerque AF, Vendemiatti JAS, Morales DA, Ormond AB, Freeman HS, Zocolo GJ, Zanoni MVB, Umbuzeiro GA:* **Chlorine disinfection of dye wastewater: Implications for a commercial azo dye mixture.** *Sci Total Environ 2013,* **442:***302–309.*

12. *OECD (Organization for Economic Co-operation and Development): Harmonized Integrated Hazard Classification System for Chemical Substances and Mixtures. Paris: OECD; 2001.*

13. *Carneiro PA, Umbuzeiro GA, Oliveira DP, Zanoni MVB:* **Assessment of water contamination caused by a mutagenic textile effluent/dyehouse effluent bearing disperse dyes.** *J Hazard Mater 2010,* **174:***694–699.*

14. *Calevro F, Filipi Deri P, Albertosi C, Batistoni R:* **Toxic effects of aluminum, chromium and cadmium in intact and regenerating freshwater planarians.** *Chemosphere 1998,* **37:***651–659.*

15. *Preza DLC, Smith D:* **Use of newborn *Girardia tigrina* (Girard, 1850) in acute toxicity tests.** *Ecotoxicol Environ Safety 2001,* **50:***1–3.*

16. *Horvat T, Kalafatić M, Kopjar N, Kavačević G:* **Toxicity testing of herbicide norflurazon on an aquatic bioindicator species - the planarians *Polycelis feline* (Daly).** *Aquat Toxicol 2005,* **73:***342–352.*

17. *Barros GS, Angelis DF, Furlan LT, Corrêa-Junior B:* **Use of freshwater planarians Dugesia (Girardia) tigrina testing toxicity of a petroleum refinery wastewater.** *J Braz Soc Ecotoxicol 2006,* **1:***67–70 (in Portuguese).*

18. *Kalafatić M, Kopjar N, Besendorfer V:* **The impairments of neoblast division in regenerating planarian *Polycelis felina* (Daly.) caused by in vitro treatment with cadmium sulfate.** *Toxicol in Vitro 2004,* **19:***99–107.*

19. *Kalafatić M, Kovačević G, Franjević D:* **Resistance of two planarians species to UV-irradiation.** *Folia Biol-Prague 2006,* **54:***103–108.*

20. *Knakievicz T, Ferreira HB:* **Evaluation of copper effects upon *Girardia tigrina* freshwater planarians based on a set of biomarkers.***Chemosphere 2008,* **71:***419–428.*

21. *Mei-Hui L:* **Effects of nonionic and ionic surfactants on survival, oxidative stress, and cholinesterase activity of planarian.***Chemosphere 2008,* **70:***1796–1803.*

22. *Knakievicz T, Vieira SM, Erdtmann B, Ferreira HB:* **Reproductive modes and life cycles of freshwater planarians (Platyhelminthes, Tricladida, Paludicula) from southern Brazil.** *Invertebr Biol 2006,* **125:***212–221.*

23. *Fukushima M, Funabiki I, Hashizume T, Osada K, Yoshida W, Ishida S:* **Detection and changes in levels of testosterone during spermatogenesis in the freshwater planarian *Bdellocephala brunnea*** . *Zool Sci 2008,* **25:***760–765.*

24. *Newmark PA, Sánchez-Alvarado A:* **Not your father's planarian: a classic model enters the era of functional genomics.** *Nat Rev Genet 2002,* **3:***210–219.*

25. *Buttarelli FR, Pellicano C, Pontieri FE:* **Neuropharmacology and behavior in planarians: translations to mammals.** *Comp Biochem Physiol C Toxicol Pharmacol 2008,* **147***(Suppl 4):399–408.*

26. *Franjević D, Krajna A, Kalafatić M, Ljubešić N:* **The effects of zinc upon the survival and regeneration of planarian *Polycelis feline*** .*Biologia 2000,* **55:***689–694.*

27. *Kato C, Mihashi K, Ishida S:* **Motility recovery during the process of regeneration in freshwater planarians.** *Behav Brain Res 2004,***150:***9–14.*

28. *Iannaconne J, Tejada M:* **Employment of regeneration of freshwater planarian *Girardia festae* (Borelli, 1898) (Tricladida: Dugesiidae) to evaluate toxicity of Carbofuran.** *Neotrop Helminthol 2007,* **1:***7–14.*

29. *Medvedev IV:* **Regeneration in two freshwater planarian species exposed to methylmercury compounds.** *Russian J Devel Biol2008,* **39:***232–235.*

30. *Novikov VV, Shieman IM, Fesenko EE:* **Effect of weak static and low-frequency alternation magnetic fields on the fission and regeneration of the planarian *Dugesia (Girardia) tigrina*** . *Bioelectromagnetics 2008,* **29:***387–393.*

31. *Reddien PW, Sánchez-Alvarado A:* **Fundamentals of planarian regeneration.** *Annu Rev Cell Dev Biol 2004,* **20:***725–757.*

32. *Saló E:* **The power of regeneration and the stem-cell kingdom: freshwater planarians (Platyhelminthes).** *Biogeosciences 2006,***28:***546–559.*

33. *Guecheva T, Henriques JAP, Erdtmann B:* **Genotoxic effects of copper sulphate in freshwater planarian in vivo, studied with the single-cell gel test (comet assay).** *Mutat Res 2001,* **497:***19–27.*

34. *Gonzáles-Estévez C, Saló E:* **Autophagy and apoptosis in planarians.** *Apoptosis 2010,* **15***:279–292.*

35. *Aboobaker AA:* **Planarian stem cell: a simple paradigm for regeneration.** *Trends Cell Biol 2011,* **21***(Suppl 5):304–311.*

36. *Piontek M:* **Application of *Dugesia tigrina* Girard in toxicological studies of aquatic environments.** *Pol Arch Hydrobiol 1998,***45***(Suppl 4):565–572.*

37. *Sáfadi RS:* **The use of freshwater planarians in acute toxicity tests with heavy metals.** *Verh Internat Limnol 1998,* **26***:2391–2392.*

38. *Orr PL, Hart DR, Craig GR: Guidance document on control of toxicity test precision using reference toxicants. Ottawa: Environmental Canada; 1990.*

39. *He L, Freeman HS, Lu L, Zhang S:* **Spectroscopic study of anthraquinone dye/amphiphile systems in binary aqueous/organic solvent mixtures.** *Dyes Pigments 2011,* **91***:389–395.*

40. *Kolasa J, Tyler S:* **Flatworms: Tuberllarians and Nemertea.** *In Ecology and classification of North American freshwater invertebrates. 2nd edition. Edited by Thorp JH, Covich AP. New York: AP; 2011.*

41. *Žnidaric D, Kalafatic M, Kopjar N:* **The survival of *Hydra oligactis* Pallas in unpleasant conditions.** *Zeitsch Angewan Zool 1995,***81***:157–163.*

42. *Raffa RB, Desai P:* **Description and quantification of cocaine withdrawal signs in Planaria.** *Brain Res 2005,* **1032***:200–201.*

43. *Villar D, Schaeffer DJ:* **Morphogenetic action of neurotransmitters on regenerating planarians - a review.** *Biomed Environ Sci 1993,***6***(Suppl 4):327–347.*

44. *Baguñà J:* **The planarian neoblast: the rambling history of its origin and some current black boxes.** *Int J Dev Biol 2012,* **56***(Suppl1-3):19–37.*

45. *Cebrià F, Guo T, Jopek J, Newmark PA:* **Regeneration and maintenance of the planarian midline is regulated by a slit orthologue.***Dev Biol 2007,* **307***:394–406.*

46. *Miyashita H, Nakagawa H, Kobayashi K, Hoshi M, Matsumoto M:* **Effects of 17β-estradiol and bisphenol A on the formation of reproductive organs in planarians.** *Biol Bull 2011,* **220***:47–56.*

47. *Vara DC, Leal-Zanchet AM, Lizardo-Daudt HM:* **Enbryonic development of *Girardia tigrina* (Girard, 1850) (Platyhelminthes, Tricladida, Paludicola).** *Bras J Biol 2008,* **68***(Suppl 4):889–895.*

48. *Benazzi M, Gremigni V:* **Developmental biology of triclad turberllarians (Planaria).** *In Developmental Biology of Freshwater Invertebrates. Edited by Harrison FW, Cowden RR. New York: Alan R Liss Inc; 1982:121–151.*

49. *Dolci-Palma IA:* **Reproduction, eating and reaction to food in planarians under different maintenance conditions.** *Psicol USP 1995,***6***(Suppl 1):173–196 (in Portuguese).*

50. *Lima OS:* **Total preparation of planarians and observations on digestion.** *Sci Cult 1954,* **6:***162–164 (in Portuguese).*

51. *Hamilton MA, Russo RC, Thurston RV:* **Trimmed Spearman-Karber method for estimating median lethal concentrations in toxicity bioassays.** *Environ Sci Technol 1977,* **11:***714–719.*

Chapter 11

Biodegradation and Decolourization of Textile Dye Wastewater Using *Ganoderma Lucidum*

Sathian Selvakumar* • Rajasimman Manivasagan • Karthikeyan Chinnappan

Environmental Engineering Laboratory, Department of Chemical Engineering, Annamalai University, Annamalai Nagar 608 002, Tamil Nadu, India

ABSTRACT

In this work, treatment of textile dye wastewater was carried in a batch reactor using *Ganoderma lucidum*. The characteristics of textile dye wastewater were studied. The effect of process parameters like pH, temperature, agitation speed and dye wastewater concentration on dye decolourization and degradation were studied. These parameters were optimized using response surface methodology (RSM). From the results, the optimized conditions were: pH 6.6, temperature 26.5 °C, agitation speed 200 rpm and dye wastewater concentration 1:2. At these optimized conditions, the maximum decolourization and COD reduction were found to be 81.4 and 90.3 %. Kinetic studies were carried out using different models like first-order, diffusional and Singh model. From the results, it was found that the degradation follows the first-order reaction model.

Keywords: Optimization RSM *Ganoderma lucidum* Dye wastewater COD

INTRODUCTION

Synthetic dyes are widely used in several industries such as textile, paper, printing, cosmetics, pharmaceuticals, colour photography and petroleum (Marmion 1991). Dyes are classified into acidic, basic, disperse, azo, diazo, anthraquinone and metal complex based on their structure. On the basis of the dyeing process, textile dyes are classified as reactive, direct, disperse, acid, basic and vat dyes (Campos et al.2001). Textile industries utilize large amounts of water during processing and also generate substantial amounts of wastewater (Hutton 1972). About 10–15 % of the dyes are lost in the wastewater during the dyeing process (Zollinger 1987). Coloured wastewater from the textile industries is one of the most obvious indicators of water pollution. Coloured dye wastewater causes severe effects on aquatic environment even in small amounts. Apart from the colour, the dischargeable dye wastewater also contains other pollutants like degradable organics, nutrients, pH altering agent, salts, sulphur, toxicants and refractory organics (Somasiri et al. 2008; Haroun and Idris 2009).

In general, physical, chemical and biological methods are used to treat the textile industry wastewater. Physical and chemical methods include adsorption, chemical precipitation, flocculation, photolysis, chemical oxidation and reduction, electro-chemical treatment and ion-pair extraction (Azmi et al. 1998; Moreria et al.2000; Rajeshkannan et al. 2010, 2011). These methods are mostly ineffective, expensive, produce side reactions, high sludge and by-products, not suited to degrade all dyes, etc. (Krull et al. 1998; Verma and Madamwar 2003). Hence, researchers have focused on biological treatment as the best alternative. The operational cost is relatively low when compared with conventional technologies (Arutchelvan et al. 2003; Jadhav and Govindwar 2006). Many microorganisms, including bacteria, fungi and actinomycetes, have been reported for their ability to decolourize dyes (Chang et al. 2001; Khehra et al. 2005). Among these microorganisms, white rot fungi are the most intensively studied dye decolourizing microbes. These fungi produce large quantities of extracellular enzymes that help to remove dyes from industrial effluent and also have the ability to resist unfavourable environmental conditions (Pointing 2001; D'Souza et al. 2006).

In this study, a white rot fungal strain, *Ganoderma lucidum*, was examined for its ability to decolourize the textile dye in industry wastewater. The effect of process variables on textile dye industry wastewater degradation was studied and optimized using response surface methodology (RSM).

MATERIALS AND METHODS

Textile dye wastewater

The textile dye wastewater was collected from a private small-scale industry located at Erode, Tamilnadu, India. The wastewater was analysed for various parameters as per the procedure given in APHA (1999), as given in Table 1. The wastewater was stored at 4 ± 1 °C in airtight plastic containers.

Table 1. Characteristics of textile dye industry wastewater

Parameters[a]	Values
pH	7.7–7.9
Colour	Brown
Total suspended solids	510
Total dissolved solids	3,880
BOD	1,132
COD	2,450
Sulphates	232
Chlorides	1,465

[a]All values except pH and colour are in mg/L

Ganoderma lucidum (MTCC- 1039) is a stock of the Microbial Type Culture Collection Centre (MTCC), Chandigarh, India. It is well preserved in the laboratory. The strain is maintained on solid medium at 4 °C. The media composition and process conditions were: agar 20 g/l; malt extract 20 g/l; temperature 25 °C; pH 6.5; incubation time 10 days.

Response surface methodology (RSM)

The RSM has several classes of designs, with its own properties and characteristics. Central composite design (CCD), Box–Behnken design and three-level factorial design are the most popular designs applied by the researchers. The Box–Behnken design was used to study the effects of the variables towards their responses and subsequently in the optimization studies. This method is suitable for fitting a quadratic surface and helps to optimize the effective parameters with a minimum number of experiments, as well as to analyse the interaction between the parameters. In order to determine the existence of a relationship between the factors and the response variables, the data collected were analysed in a statistical manner using regression. A regression design is normally employed to model a response as a mathematical function (either known or empirical) of a few continuous factors and good model parameter estimates are desired.

The coded values of the process parameters are determined by the following equation

$$x_i = \frac{X_i - X_0}{\Delta X} \quad (1)$$

where x_i is the coded value of the ith variable, X_i the uncoded value of the ith test variable and X_0 the uncoded value of the ith test variable at the centre point.

Regression analysis was performed to estimate the response function as a second-order polynomial

$$Y = \beta_0 + \sum_{i=1}^{k} \beta_i X_i + \sum_{i=1}^{k} \beta_{ii} X_i^2 + \sum_{i=1, i<j}^{k-1} \sum_{j=2}^{k} \beta_{ij} X_i X_j \quad (2)$$

where Y is the predicted response and β_i, β_j, β_{ij} are the coefficients estimated from regression. They represent the linear, quadratic and cross products of x_1, x_2, x_3 on response.

The regression and graphical analysis with statistical significance were carried out using Design-Expert software (version 7.1.5, Stat-Ease, Inc., Minneapolis, USA). To visualize the relationship between the experimental variables and responses, the response surface and contour plots were generated from the models. The optimum values of the process variables were obtained from the response surface.

The adequacy of the models was further justified through analysis of variance (ANOVA). Lack-of-fit is a special diagnostic test for adequacy of a model that compares the pure error, based on the replicate measurements to the other lack-of-fit and model performance. F value, calculated as the ratio between the lack-of-fit mean square and the pure error mean square, is the statistic parameter used to determine whether the lack-of-fit is significant or not, at a significance level.

Experimental procedure

Experiments were carried out in a 500-ml Erlenmeyer flask, according to the Box–Behnken design given in Table 2. The process parameters chosen for this study were pH, temperature, agitation speed and initial dye wastewater concentration. The pH of the sample was adjusted to 5, 7 and 9 by adding acid or base as required. Batch studies were carried out by varying the temperature to 23, 28 and 33 °C, respectively. Agitation speed was varied to 100, 200 and 300 rpm by means of an agitator. The initial concentration of dye wastewater was varied as 1:1 and 1:2. The dye concentration was measured with Bio-

Spectrophotometer (Model: BL-200, ELICO, India) at a wavelength of 395 nm. COD of the sample was analysed using the procedure given in APHA.

Table 2. Box–Behnken design-based experimental conditions and results for textile dye wastewater treatment using*Ganoderma lucidum*

Run no.	*A*-pH	*B*-Temperature	*C*-Agitation speed	*D*-Dye wastewater concentration	% decolourization		% degradation	
					Experimental	Predicted	Experimental	Predicted
1	5	28	150	1:2	75.3	71.49	81.4	81.06
2	9	23	150	1:1	60.3	57.25	68.5	66.47
3	5	28	150	Raw	40.2	38.12	51.8	47.11
4	7	33	150	Raw	45.6	48.91	58.8	58.32
5	7	28	200	Raw	49.6	50.29	60.2	62.65
6	9	28	100	1:1	56.3	60.33	63.8	64.34
7	7	28	100	1:2	78.5	74.81	84.5	84.28
8	7	23	150	1:2	79.5	81.11	88.9	88.44
9	7	23	150	Raw	40.2	39.44	57.5	58.49
10	7	23	200	1:1	65.3	66.89	74.2	75.95
11	9	28	200	1:1	62.3	61.93	69.5	67.30
12	7	28	150	1:1	70.3	70.07	78.8	78.80
13	7	28	150	1:1	69.6	70.07	78.9	78.80
14	7	33	100	1:1	62.7	59.20	70.5	67.46
15	9	28	150	1:2	66.8	66.97	72.5	75.90
16	7	33	150	1:2	59.8	65.48	87.5	85.57
17	9	33	150	1:1	60.2	57.52	65.2	66.45
18	9	28	150	Raw	40.2	42.10	53.6	52.66
19	7	28	150	1:1	70.3	70.07	78.7	78.80
20	5	23	150	1:1	61.2	60.87	66.8	67.78
21	7	33	200	1:1	71.2	68.35	79.6	79.53
22	7	28	200	1:2	79.2	79.25	88.9	88.45
23	5	28	200	1:1	64.3	65.19	72.6	71.12
24	7	23	100	1:1	65.9	66.84	75.3	74.08
25	7	28	100	Raw	48.6	45.54	50.2	52.88
26	5	28	100	1:1	52.3	57.59	58.9	60.15
27	5	33	150	1:1	54.4	54.44	60.5	64.77
28	7	28	150	1:1	70.1	70.07	78.4	78.80
29	7	28	150	1:1	70.5	70.07	78.8	78.80
30	7	28	150	1:1	70.0	70.07	78.0	78.80

Laccase enzyme assay

The laccase activity was determined using 2, 2′-azino-bis (3-ethylbenzthiazoline-6-sulphonic acid) (ABTS) as the substrate (Rasera et al. 2009). The laccase reaction mixture contained 0.5 ml of 0.45 mM ABTS, 1.2 ml of 0.1 M phosphate buffer (pH 6.0) and 0.5 ml of filtrate to give a final reaction volume of 2.2 ml. The oxidation of the substrate (ABTS) was monitored by the increase in the absorbance at 420 nm using Shimadzu UV-1800 spectrophotometer (ELICO, India) over 90 s at 30 °C, using $\varepsilon = 3.6 \times 104\ cm^{-1}\ M^{-1}$. Enzymatic activity was expressed as 1U = 1 µmol of ABTS oxidized per min at 25 °C (Vinothkumar et al.2011).

RESULTS AND DISCUSSION

The effect of process parameters (independent variables) on the decolourization and degradation of textile dye wastewater was studied. The second-order polynomial coefficients for each term of the equation Eqs. (3), (4) were determined using the Design-Expert 7.1.5. The experimental and predicted values of percentage decolourization and degradation are given in Table 2.

$$\begin{aligned} \%\ \text{Decolourization} = {} & 69.78 - 0.13A - 1.54B + 2.30C \\ & + 14.56D + 1.68AB - 1.50AC - 2.13AD \\ & + 2.28BC - 6.28BD - 0.075CD - 8.16A^2 \\ & - 4.10B^2 - 0.36C^2 - 6.95D^2 \end{aligned} \tag{3}$$

$$\begin{aligned} \%\ \text{COD reduction} = {} & 78.72 + 0.092A - 0.76B \\ & + 3.48C + 14.30D + 0.75AB - 2.00AC \\ & - 2.68AD + 2.55BC - 0.68BD - 1.40CD \\ & - 10.44A^2 - 1.91B^2 - 2.55C^2 - 4.10D^2 \end{aligned} \tag{4}$$

where *A*, *B*, *C* and *D* are the coded values of the process variables, pH, temperature (°C), agitation speed (rpm) and wastewater concentration, respectively.

The results were analysed by using analysis of variance (ANOVA) and are given in Table 3. The ANOVA of the quadratic regression model indicates that the model is significant. In this work, the model *F* value 19.79 and 34.04 for decolourization and COD reduction implies that the models are significant. The smaller the magnitude of *P*, more significant is the corresponding coefficient. *P* value less than 0.05 indicates that the model terms are significant. From the *P* values, it was found that the variables *C, D, BD,* A^2, B^2 and D^2were significant model terms for decolourization and *C, D,* A^2, C^2 and D^2 were significant model terms for COD reduction. From the ANOVA table, it was found that the linear effect of dye wastewater concentration is more significant for textile dye wastewater treatment followed by agitation speed. Also, the interactive effect of temperature and dye wastewater concentration is significant foe dye decolourization.

Table 3. ANOVA for the decolourization and COD reduction of textile dye wastewater

Source	% Decolourization			% COD reduction		
	Coefficient factor	*F*	*P* > *F*	Coefficient factor	*F*	*P* > *F*
Model	69.78	19.79	<0.0001	78.72	34.04	<0.0001
A	−0.13	0.017	0.8990	0.092	0.014	0.9074
B	−1.54	2.23	0.1572	−0.76	0.96	0.3440
C	2.30	4.97	0.0426	3.48	20.24	0.0005
D	14.56	199.24	<0.0001	14.30	341.09	<0.0001
A × *A*	−8.16	33.84	<0.0001	−10.44	98.26	<0.0001
B × *B*	−4.10	8.54	0.0112	−1.91	3.30	0.0906
C × *C*	−0.36	0.066	0.8008	−2.55	5.87	0.0295
D × *D*	−6.95	24.53	0.0002	−4.10	15.17	0.0016
A × *B*	1.68	0.88	0.3643	0.75	0.31	0.5848
A × *C*	−1.50	0.71	0.4152	−2.00	2.22	0.1581
A × *D*	−2.13	1.42	0.2540	−2.68	3.98	0.0659
B × *C*	2.28	1.62	0.2236	2.55	3.62	0.0780
B × *D*	−6.28	12.34	0.0034	−0.68	0.25	0.6226
C × *D*	−0.075	1.73E−003	0.9671	−1.40	1.09	0.3142
R^2	0.9519			0.9715		
Pred R^2	0.9038			0.9429		
Adj R^2	0.8240			0.8357		
Std. Dev.	3.5700			2.6800		
C.V	5.79 %			3.79 %		
Adeq precision	16.730			21.429		

The predicted R^2 of 0.8240 (decolourization), 0.8357 (COD reduction) was in reasonable agreement with the adjusted R^2 of 0.9038 (decolourization) and 0.9429 (COD reduction). The fit of the model is also expressed by the coefficient of regression R^2, which is found to be 0.9519 for decolourization and 0.9715 for COD reduction, indicating that more than 95 % of the variability in the response could be explained by the model. This implies that the prediction of experimental data is quite satisfactory. Adequate precision measures the signal to noise ratio. A ratio greater than 4 is desirable. In this work, the ratio was found to be 16.730 and 21.429 for decolourization and COD reduction, which indicates an adequate signal.

The magnitude of coefficient factors in Table 3 gives the positive contribution of agitation speed and wastewater concentration and negative contribution of pH and temperature on dye wastewater decolourization. For COD reduction, pH, agitation speed and dye wastewater concentration show the positive contribution, whereas temperature has a negative effect.

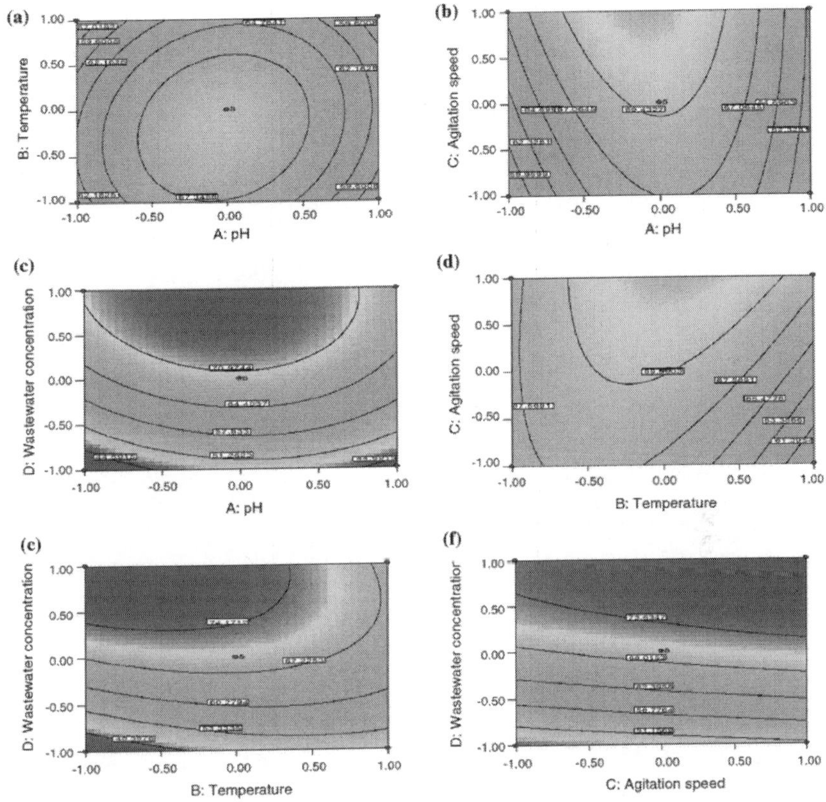

Figure 1. Contour plot showing the interactive effect of **a** pH and temperature, **b** pH and agitation speed, **c** pH and wastewater concentration, **d** temperature and agitation speed, **e** temperature and wastewater concentration, and **f** agitation speed and wastewater concentration on decolourization of textile dye wastewater

To investigate the interactive effect of two factors on the decolourization and degradation of dye wastewater, response surface methodology was used and contour plots were drawn. Response surface plots as a function of two factors at a time, maintaining all other factors at fixed levels, are more helpful in understanding both the main and the interactive effects of two factors. The response surface curves for the decolourization and degradation of textile dye wastewater are shown in Figs. 1 and 2. The nature of the response surface curves shows the interaction between the variables. The elliptical shape of the curve indicates good interaction between the two variables and circular shape indicates no interaction. From the figures, it is observed that the elliptical nature of the contour in graphs depicts the mutual interactions of all the variables. There is a relatively significant interaction between every two variables and a maximum predicted yield as indicated by the surface confined in the smallest ellipse in the contour diagrams.

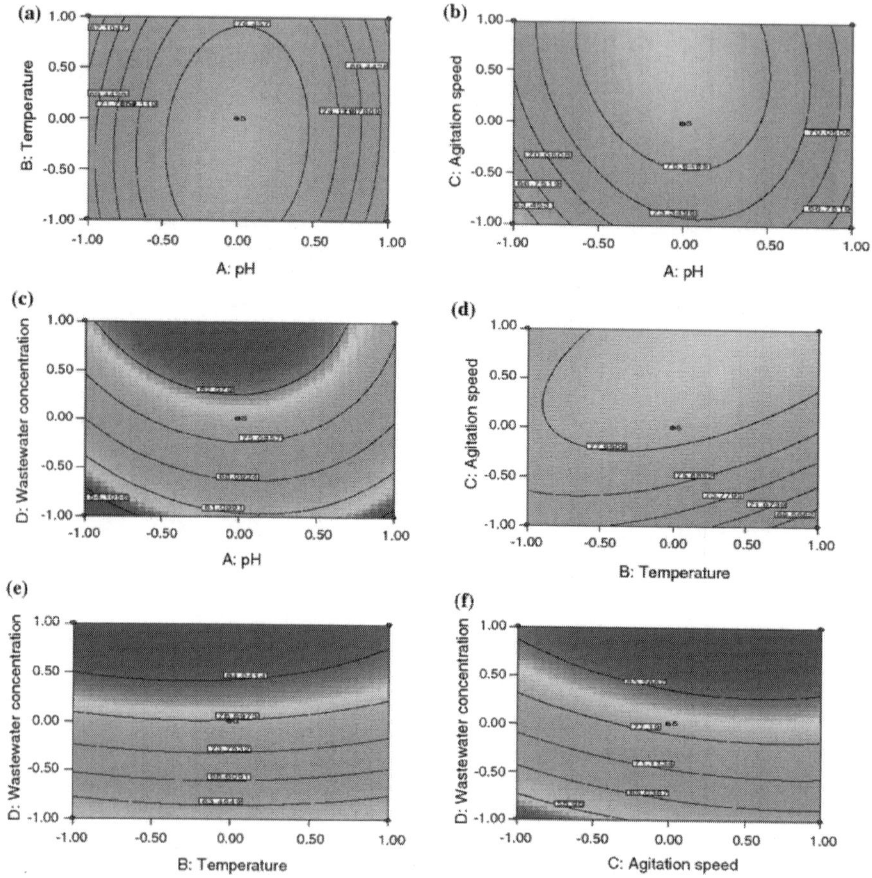

Figure 2. Contour plot showing the interactive effect of **a** pH and temperature, **b** pH and agitation speed, **c** pH and wastewater concentration, **d** temperature and agitation speed, **e** temperature and wastewater concentration and **f** agitation speed and wastewater concentration on COD reduction of textile dye wastewater

Decolourization of textile dye wastewater

Figure 1a shows the interactive effect of pH and temperature on textile dye decolourization. From the figure, it was inferred that increase in pH (up to 6.4) increases the dye decolourization efficiency. After that, the decolourization efficiency decreases. A similar trend was observed in Fig. 1b and c. The pH has a major effect on the efficiency of dye decolourization, and the optimal pH for colour removal is often between 6.0 and 7 for most of the dyes. The pH tolerance of decolourizing fungi is quite important for decolourization of textile dye wastewater.

It is clear from Fig. 1a that percentage decolourization of dye increased with an increase in temperature from 23 to 26.5 °C. The percentage decolourization of dye decreased with further increase in temperature up to 33 °C. The decolourizing activity was significantly suppressed at higher temperatures. This may be due to the loss of cell viability or deactivation of the enzymes responsible for decolourization. Hence the optimum temperature for maximum decolourization of textile dye wastewater using *Ganoderma lucidum* was 26.5 °C. From Fig. 1b, it was observed that the percentage of decolourization increased with increase in agitation speed up to 200 rpm. This shows that oxygen is needed for the microorganism to degrade the dye wastewater. A similar trend is observed in Fig. 1d, f. The decrease in initial textile dye wastewater concentration increases the decolourization. This is clearly depicted in Fig. 1c, e and f. From the figures, it is inferred that the percentage removal of dyes increases with decrease in initial dye concentration. This is because at higher concentrations, the chemicals and other pollutants present in the dye wastewater inhibit the growth of microorganism.

Degradation of textile dye wastewater

pH is one of the important factors in the treatment of textile dye wastewater by microorganism. The interactive effect of pH and temperature on biodegradation of textile wastewater using *Ganoderma lucidum* is shown in Fig. 2a. From the figure, it was observed that increase in pH up to 6.6 increased the COD reduction. Further increase in pH leads to decrease in COD reduction. The interactive effects of other parameters are shown in Fig. 2b–f. The trend observed for other parameters was similar to the decolourization profile. The optimum conditions were also the same, except agitation speed (175), for COD reduction.

The second-order polynomial equation obtained from RSM was used to find the optimum conditions. The equation was solved in MATLAB 7.0. The optimum condition for the maximum decolourization was found to be: pH 6.6, temperature 26.5 °C, agitation speed 200 rpm and dye wastewater concentration 1:2.The optimal conditions predicted using RSM has been validated using experiments. At the optimized condition, the maximum colour removal and COD reduction were found to be 81.4 and 90.3 %, respectively.

At the optimum condition, decolourization of textile wastewater was studied by analysing the supernatants at different time intervals, in a UV-spectrophotometer in the range of 300–800 nm. The results obtained are shown in Fig. 3. A peak is observed at λ_{max} 395 nm in the UV–Vis spectra. The peak decreases as the day progresses, which shows the decolourization of textile dye using *Ganoderma lucidum* at the end of the 5th day of operation. The peak at 395 nm attains nearly zero, which shows almost complete decolourization of the textile dye.

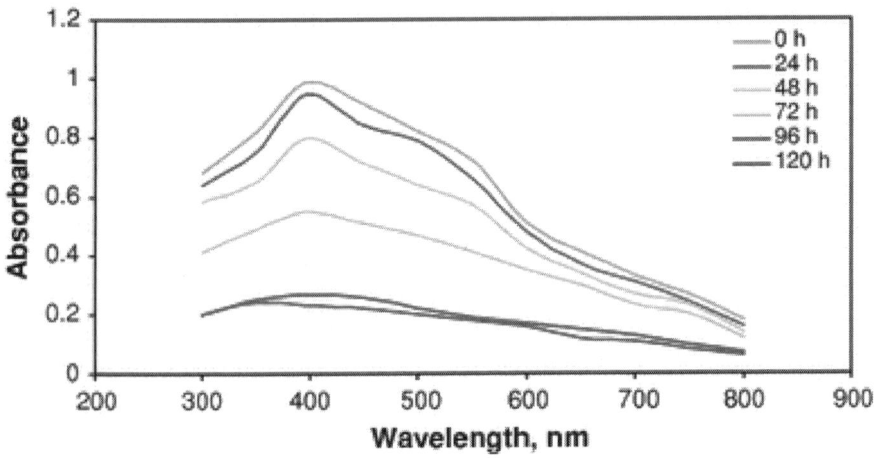

Figure 3. UV spectra of textile dye wastewater decolourization by *Ganoderma lucidum* at the optimum condition

The laccase activity was measured at these optimum conditions and is shown in Fig. 4. From the figure, it was inferred that the laccase activity increased till the 4th day and reached a maximum of 6.1 U/ml; thereafter, the laccase enzyme activity decreases. The decolourization process increases with increase in enzyme production and decreases with decrease in enzyme activity. After 6 days, no enzyme activity was found and the decolourization process stopped. This clearly shows that the decolourization process occurs by the action of the laccase enzyme.

Kinetics

In this study, first-order model, diffusional model and Singh model were tried to fit the experimental data obtained from the batch degradation of textile dye wastewater using *Ganoderma lucidum*.

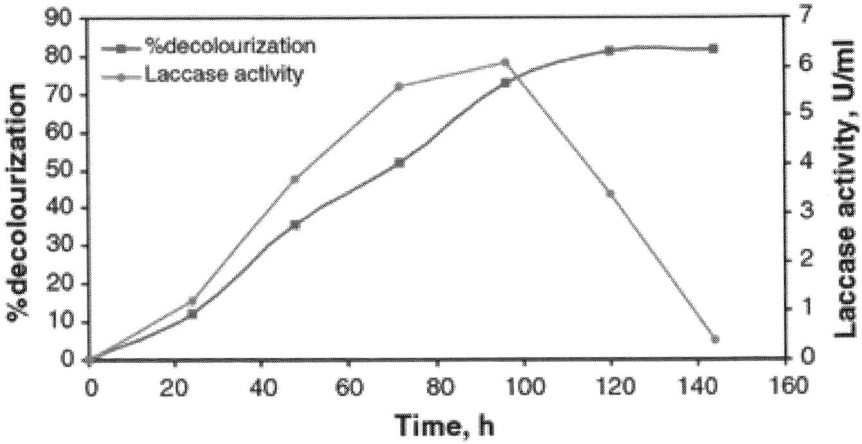

Figure 4. Laccase activity during the decolourization of textile dye wastewater by *Ganoderma lucidum*

First-order model

The First-order model is

$$-\frac{dC_s}{dt} = k_1 C_S \tag{5}$$

On integration between known limits and rearranging, the above model becomes

$$\ln\left(\frac{C_{so}}{C_s}\right) = k_1 t \tag{6}$$

where C_{so} is the initial substrate concentration in mg COD/L, C_s is the substrate concentration in mg COD/L, t is the degradation time in h, k_1 is the first-order rate constant in h^{-1}.

Diffusional model

The diffusional model is given by

$$-\frac{dC_s}{dt} = k_2 C_s^{0.5} \tag{7}$$

when integrated between the known limits, the above equation becomes

$$\sqrt{C_s} - \sqrt{C_{so}} = \frac{k_2}{2} t \tag{8}$$

where k_2 = rate constant for diffusional model.

Singh model

The Singh model is given by

$$-\frac{dC_s}{dt} = \frac{k_3 C_s}{1+t} \tag{9}$$

Integrating the above equation between the proper limits, it becomes

$$\ln\left(\frac{C_s}{C_{so}}\right) = -k_3 \ln(1+t) \tag{10}$$

where k_3 is the rate constant for the Singh model.

The data obtained from the batch study were fitted to the first-order, diffusional and Singh model and are shown in Figs. 5a–c. The rate constants, k_1, k_2 and k_3 were calculated from the slope of the straight line by the least square (LSQ) fit in the figures. The values of k_1, k_2 and k_3 were 0.0251, −0.3597 and 0.4197, respectively. The determination coefficient (R^2) for first-order, diffusional and Singh model was 0.9333, 0.9383 and 0.5115, respectively. The high R^2 value for the first order indicates the fitness of the model for the degradation of textile dye wastewater. The negative value of the rate constant for diffusional model and low R^2 value for Singh models show the inability of these models to describe this process.

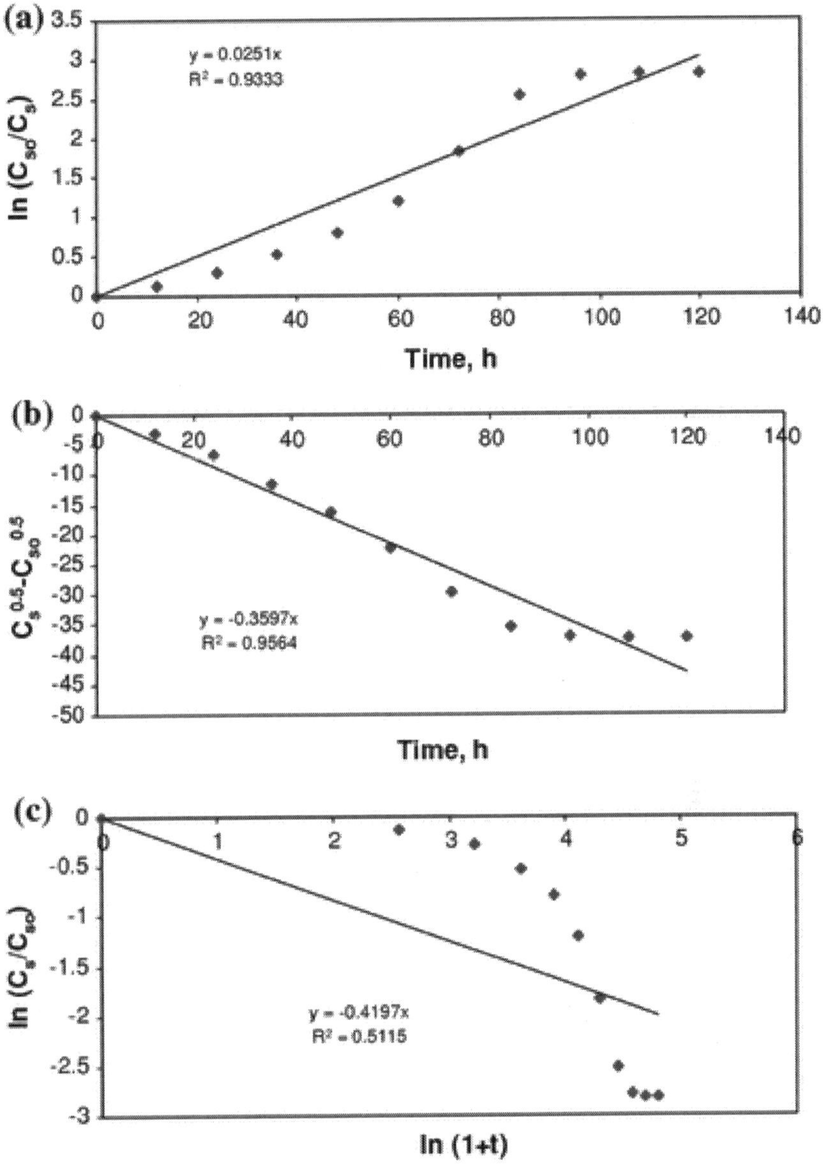

Figure 5. Kinetic plots for degradation of textile dye wastewater. **a** First-order model, **b** diffusional model and **c** Singh model

CONCLUSIONS

In this study, a white rot fungi, *Ganoderma lucidum*, was utilized to treat the textile dye wastewater. RSM was applied to optimize the process parameters. From the results, it was found that a maximum of 81.4 % colour removal and 91.3 % COD reduction occurs at the optimized condition. The UV spectrum confirms the decolourization. Various models were tried to study the kinetics of textile dye degradation. From the results, it was found that the degradation of textile dye wastewater follows first-order kinetics. Hence, it was concluded that *Ganoderma lucidum* could be utilized for the effective treatment of textile dye wastewater.

ACKNOWLEDGMENTS

The authors wish to express their gratitude for the support extended by the authorities of Annamalai University, Annamalai Nagar, India, in carrying out the research work in the Environmental Engineering Laboratory, Department of Chemical Engineering.

CONFLICT OF INTEREST

There is no conflict of interest with the organization which sponsored this work.

OPEN ACCESS

REFERENCES

1. APHA (1999) Standard methods for the examination of water and wastewater, 20th edn. American Public Health Association (APHA), Washington DC

2. Arutchelvan V, Albino DJ, Muralikaishnan V, Nagarajan S (2003) Decolourization of textile mill effluent by*Sporotrichum pulverulentum*. Indian J Environ Ecoplaning 7:59–62

3. Azmi W, Sani RJ, Banerjee UC (1998) Biodegradation of triphenylmethane dyes. Enzyme Micro Technol 22:185–191

4. Campos R, Kandelbauer A, Robra KH, Artur Cavaco PauloGubitz GM (2001) Indigo degradation with purified laccases from *Trametes hirsuta* and *Sclerotim rolfsii*. J Biotechnol 8:131–139

5. Chang JS, Chou C, Chen SY (2001) Decolorization of azo dyes with immobilized *Pseudomonas luteola*. Process Biochem 36:757–763

6. D'Souza DT, Tiwari R, Sah AK, Raghukumar C (2006) Enhanced production of laccase by a marine fungus during treatment of colored effluents and synthetic dye. Enzyme Micro Technol 38:504–511

7. Haroun M, Idris A (2009) Treatment of textile dye with an anaerobic fluidized bed reactor. Desalination 237:357–366

8. Hutton DG (1972) Improved biological wastewater treatment. Dupoint Innovation 2:6–26

9. Jadhav JP, Govindwar SP (2006) Biotransformation of malachite green by *Saccharomyces cerevisiae*. Yeast 23:315–323

10. Khehra MS, Saini HS, Sharma DK, Chadha BS, Chimmi SS (2005) Decolorization of various azo dyes by bacterial consortium. Dyes Pig 67:55–61

11. Krull R, Hemmi M, Otto P, Hempel DC (1998) Combined biological and chemical treatment of highly concentrated residual dyehouse liquors. Water Sci Technol 38:339–346

12. Marmion DM (1991) Handbook of US colorants. Foods, drugs, cosmetics and medical devices, 3rd edn. Wiley, New York

13. Moreria MT, Mielgo I, Feijoo G, Lema JM (2000) Evaluation of different fungi strains in the decolourisation of synthetic dyes. Biotechnol Lett 22:499–503

14. Pointing SB (2001) Feasibility of bioremediation by white-rot fungi. Appl Microbiol Biot 57:20–33

15. Rajeshkannan R, Rajasimman M, Rajamohan N (2010) Optimization, equilibrium and kinetics studies on sorption of Acid Blue 9 using brown marine algae *Turbinaria conoides*. Biodegradation 21:713–727

16. Rajeshkannan R, Rajasimman M, Rajamohan N (2011) Decolourization of malachite green using tamarind seed: optimization, isotherm and kinetic studies. CICEQ 17:67–79

17. Rasera K, Ferla J, Dillon AJP, Riveiros R, Zeni M (2009) Immobilization of laccase from *Pleurotus sajor-caju* in polyamide membranes. Desalin. 245:657–661

18. Somasiri W, Li X, Ruan W, Jian C (2008) Evolution of efficacy of upflow anaerobic sludge blanket reactor in removal of colour and reduction of COD in real textile wastewater. Bioresour Technol 99:3692–3699

19. Verma P, Madamwar D (2003) Decolorization of synthetic dyes by a newly isolated strain of *Serratia marcescens*. World J Microbiol Biotechnol 19:615–618

20. Vinothkumar V, Kirupha SD, Periyaraman P, Sivanesan S (2011) Screening and induction of laccase activity in fungal species and its application in dye decolorization. Afr J Microbiol Res 5:1261–1267

21. Zollinger H (1987) Color chemistry-synthesis, properties and application for organic dyes and pigments. VCH Publications, NY, pp 92–102

Index